A Methodology for Modeling the Flow of Military Personnel Across Air Force Active and Reserve Components

Lisa M. Harrington, James H. Bigelow, Alexander Rothenberg, James Pita, Paul D. Emslie

RAND Project AIR FORCE

Prepared for the United States Air Force
Approved for public release; distribution unlimited

T0308364

For more information on this publication, visit www.rand.org/t/RR825

Library of Congress Cataloging-in-Publication Data is available for this publication.

ISBN: 978-0-8330-9302-8

Published by the RAND Corporation, Santa Monica, Calif.

© Copyright 2016 RAND Corporation

RAND® is a registered trademark.

Support RAND
Make a tax-deductible charitable contribution at
www.rand.org/giving/contribute

www.rand.org

Preface

The Air Force is placing increased emphasis on managing its airmen as a total force. Yet many aspects of personnel management are conducted within the confines of a particular component—whether active, guard, or reserve. As a result, when personnel policies are implemented in one component, little is known or considered about the effect that those policies might have on personnel flows into and out of other components. The degree to which this is important varies by career field, so examination of such concerns must be conducted not only at an aggregate level but also for individual specialties.

Total force personnel management requires tools that provide managers with insight on personnel flows across components and how those flows are affected by personnel policies that lead to changes in accessions, retention, affiliation, and retirements. With a view toward shaping the future force size and mix from a total force perspective, the Assistant Secretary of the Air Force for Manpower and Reserve Affairs (SAF/MR) asked RAND Project AIR FORCE (PAF) to help improve Air Force capability to analyze and capitalize on military personnel flows across the total force. In response to this request, RAND developed a component flow model described in this report. The model's description and capabilities will be of interest to manpower and personnel managers and analysts both in and outside the Air Force.

The research reported here was sponsored by the Assistant Secretary of the Air Force for Manpower and Reserve Affairs and conducted within the Manpower, Personnel, and Training Program of RAND Project AIR FORCE.

RAND Project AIR FORCE

RAND Project AIR FORCE, a division of the RAND Corporation, is the U.S. Air Force's federally funded research and development center for studies and analysis. PAF provides the Air Force with independent analyses of policy alternatives affecting the development, employment, combat readiness, and support of current and future aerospace forces. Research is conducted in four programs: Force Modernization and Employment; Manpower, Personnel, and Training; Resource Management; and Strategy and Doctrine. The research reported here was prepared under contract FA7014-06-C-0001.

Additional information about PAF is available on our website:
http://www.rand.org/paf

Table of Contents

Figures

Tables

Summary

Total force management is receiving renewed emphasis across the Department of Defense (DoD), as more than a decade of war in Iraq and Afghanistan draws to a close and government-wide fiscal pressures limit the availability of resources. For the Air Force, the question at hand is how to best make use of the total force—the active component, the Air Force Reserve, and the Air National Guard—to meet future mission requirements in a sustainable way. Integral to managing an effective and sustainable force mix is the need to invest in accessing, developing, and retaining the right mix of active and reserve component members.

While the Air Force has a long history of operational integration, with its reserve components (guard and reserve) trained to the same levels of readiness as its active component, a new level of integration is now needed in force management. For the most part, force management today occurs separately in the individual components, which means that the effect of policy changes in one component on personnel and force structure in others is not well understood. What is needed is an approach to force management that considers the effects of personnel policies on all components simultaneously.

Force planners require tools to help them answer questions concerning the feasibility of sustaining the number and mix of personnel necessary to meet mission requirements. Total force personnel managers require tools that provide insight on personnel flows across components and how those flows are affected by changes in accessions, retention, affiliation, and retirements, for example. They need tools that can help them understand whether policy changes are necessary to meet personnel goals, not only in the aggregate but also within career fields and for individual members.

Given the need for more integrated human capital management, the Assistant Secretary of the Air Force for Manpower and Reserve Affairs (SAF/MR) asked RAND PAF to help improve Air Force capability to analyze and capitalize on military personnel flows across the total force. In response, RAND built the *Total Force Flow Model*.

The foundation of the model is career history data sets that cover active, guard, and reserve component personnel. The career history documents the behaviors of airmen as they enter the Air Force and proceed along their career path, during which they may move from one component to another (for example, enter the active component and subsequently leave to join a reserve component), change career fields, or return to civilian life at any time during that career. Building these career histories required data elements such as component, full- or part-time status, grade, years of service, specialty, date of entry, source of commission or enlistment, separations from military service, and transfers from one component to another. Collecting data of this type for each year of military service (including active and reserve) and using it to create individual career profiles for personnel across all three Air Force components has never been done in such a comprehensive manner. The uniqueness of the data sets produced and of the methodology used to create them makes this study a valuable contribution to future total force personnel management.

The model itself builds from existing total force management tools such as RAND's Total Force Blue Line model and sustainment and utilization tools. These existing capabilities are useful for force management and contribute to an improved picture of personnel flows across the total force. But these tools have limitations—they do not cover all career fields in all three components, and they do not have the capability to forecast personnel behavior into the future. Future personnel managers will need additional capabilities to manage the total force. These include the ability to

- forecast future inventories, including the effects of changes in economic conditions
- establish goals for personnel policy changes in a particular component or across the total force and examine the resulting changes in personnel flows
- examine the interrelated effects of changes to accessions, affiliations, separations, and retirements
- assess actions necessary to more efficiently meet component contributions to the overall requirement.

RAND's new component flow model adds these capabilities to the suite of existing tools. Using the personnel career histories and historical manpower data, the model can be used for two primary purposes. The first is to plot and analyze historic data trends. The second is to forecast personnel flows in response to various economic conditions or changes in personnel behavior (accessions, affiliations, separations, and retirements) and to evaluate and optimize the ability to match inventory to requirements based on different assumptions about manpower requirements. We envision this model being used by personnel managers to develop policy objectives in response to proposed changes to organization and mission, active and reserve component mix, or other proposed changes to total force personnel flows and inventories.

This report describes the methodology used to identify and organize historical data to build the career profiles and associated manpower data sets and describes in detail the capabilities of the Total Force Flow Model.

Acknowledgments

We are grateful to many people involved in this research. In particular, we would like to thank our Air Force sponsors, initially Sheila Earle (SAF/MR), William Booth (SAF/MRM), and later Dan Sitterly (acting SAF/MR). In addition, we would like to thank several action officers for their help and guidance throughout this study: Todd Remington (SAF/MRR), Greg Parton (AF/A1MR), and Gene Blinn (SAF/A1MR). This research would not have been possible without their contributions.

Many RAND colleagues contributed to this effort. Most importantly, the contributions of Ray Conley, initially as principal investigator and then as program director, were key. The authors thank John Boon, Judith Mele, and Paul Emslie for their efforts in preparing analytic data sets for use in the models; Michael McGee for his work on force utilization; Ronald McGarvey for his early work on the optimization component of the model; Eric Jorgensen and Barbara Bicksler for integrating the document and adding many contextual references; and Samantha Bennett for editing this document. This research benefited from helpful insights and comments provided by our RAND colleague Al Robbert. We also want to thank our reviewers, James Hosek and Pete Schirmer, for their thoughtful comments that greatly improved this report.

Chapter One. Introduction

> Emphasis will be given to concurrent consideration of the total forces, active and reserve, to determine the most advantageous mix to support national strategy and meet the threat. A total force concept will be applied to all aspects of planning, programming, manning, equipping and employing the Guard and Reserve Forces.
>
> Secretary of Defense Melvin Laird, 1970[1]

Even though the total force concept is nearly half a century old, it is receiving renewed emphasis across the Department of Defense (DoD), as more than a decade of warfighting in Iraq and Afghanistan draws to a close and government-wide fiscal pressures limit the availability of resources. For the Air Force, and the rest of the military services in DoD, the question at hand is how to best make use of the total force—the active component,[2] the Air Force Reserve, and the Air National Guard—to meet future mission requirements in a sustainable way.

The active and reserve component mix of each military service directly affects its total force's capability, capacity, efficiency, and cost-effectiveness. Integral to managing an effective and sustainable force mix, though not always explicitly addressed during planning, is the need to invest in accessing, developing, and retaining the right mix of active and reserve component members. For the Air Force, this means having the right amount of the right kind of airmen in the right place at the right time.

In January 2013, Secretary of the Air Force Michael Donley and Chief of Staff General Mark Welsh created the Total Force Task Force to look closely at total force integration. Chaired by three major generals, one from each component, the task force "conducted a comprehensive review of total force requirements, offered many ideas for improving collaboration between the three components, and presented a starting point for future total force analysis and assessment efforts."[3] That work is being continued by a transitional organization called the Total Force

---

[1] Martin Binkin, *Who Will Fight the Next War? The Changing Face of the American Military*, Washington, D.C.: The Brookings Institution, 1993. In August 1973, the Secretary of Defense declared that the "Total Force is no longer a 'concept.' It is now the Total Force Policy which integrates the active, Guard, and Reserve forces into a homogeneous whole." For more discussion, see Bernard D. Rostker Charles Robert Roll, Jr., Marney Peet, Marygail K. Brauner, Harry J. Thie, Roger Allen Brown, Glenn A. Gotz, Steve Drezner, Bruce W. Don, Ken Watman, Michael G. Shanley, Fred L. Frostic, Colin O. Halvorson, Norman T. O'Meara, Jeanne M. Jarvaise, Robert Howe, David A. Shlapak, William Schwabe, Adele R. Palmer, James H. Bigelow, Joseph G. Bolten, Deena Dizengoff, Jennifer H. Kawata, Hugh G. Massey, Robert Petruschell, S. Craig Moore, Thomas F. Lippiatt, Ronald E. Sortor, J. Michael Polich, David W. Grissmer, Sheila Nataraj Kirby, and Richard Buddin, *Assessing the Structure and Mix of Future Active and Reserve Forces*, Santa Monica, Calif.: RAND Corporation, MR-140-1-OSD, 1992.

[2] Section 8075 of Title 10 of the U.S. Code calls this component the Regular Air Force, but it is more generally referred to as the active component.

[3] Deborah Lee James and Mark A. Welsh III, Written Statement to the Senate, Committee on Armed Services, *The National Commission on the Structure of the Air Force*, April 29, 2014, p. 4.

Continuum, the purpose of which is to facilitate development of a permanent staff structure focused on the same issues.[4]

The Total Force Task Force and Total Force Continuum provided assistance to the National Commission on the Structure of the Air Force, mandated in the National Defense Authorization Act for Fiscal Year 2013, to address congressional concerns about the service's Fiscal Year 2013 budget proposal. The commission's findings, delivered in January 2014, embrace greater integration of the Air Force's three components, arguing that this will give reserve component airmen more opportunities to serve, which, in turn, can reduce total force military personnel costs and increase funds available for readiness, modernization, and recapitalization.[5]

Greater levels of integration as recommended by the commission and being considered by the Air Force will require a commensurate level of integration in force management—something that does not currently exist to the degree needed. In fact, management of the Air Force active, guard, and reserve manpower and personnel often occurs only within the individual component, with little clear understanding of how policy changes in one component might impact the personnel and force structure in the others. What is needed is an approach to force management that considers the effects of changes to manpower targets and personnel flows on all components simultaneously.

Thus, force planners will need new and improved tools to help them analyze proposed mission shifts and the associated transition effects. They will need tools that will help them answer questions about the feasibility of sustaining the number and mix of personnel necessary to meet mission requirements. And they will need to understand the effect on personnel flows of changes in accession, retention, affiliation, and retirements—not only in the aggregate but also within career fields and for individual members.

The distribution of Air Force military manpower results from the flow of personnel into and out of the active and reserve components, as shown in Figure 1.1. The active, guard, and reserve components all access civilians with no prior military service (nonprior service or NPS), and personnel leave military service to go back to the civilian sector, sometimes permanently and other times only temporarily. But personnel also transition between components via direct affiliation or after temporarily separating from and later returning to the military (a "break in service"), sometimes more than once in a career—moving from the active component to one of the reserve components, moving from a reserve component to the active component, or moving between the guard and reserve. When force planning occurs only within a single component, it fails to account for and properly capitalize on these affiliation flows.

---

[4] James and Welsh, 2014.

[5] National Commission on the Structure of the Air Force, *Report to the President and Congress of the United States*, Arlington, Va., January 30, 2014, p. 11.

Figure 1.1. Total Force Personnel Flows

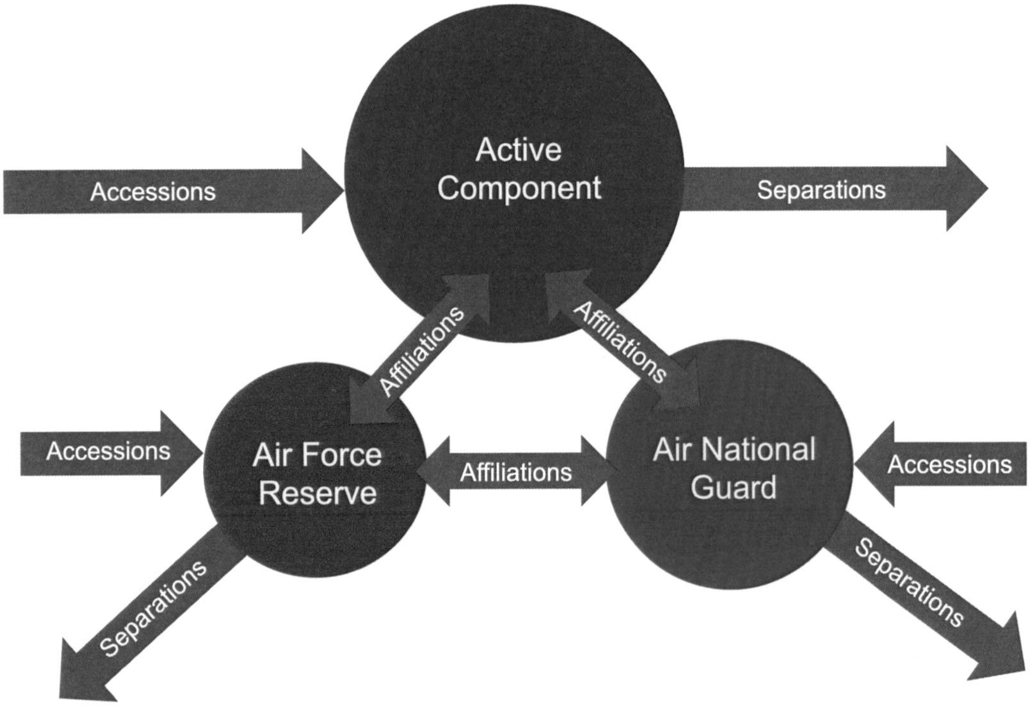

NOTE: Accessions include nonprior and prior service entrants.

Military personnel flows reflect a complex set of relationships that vary over time. Using active component pilots as an example, Figure 1.2 illustrates the flows of individuals into the inventory (top half of graph) and out of the inventory (bottom half of graph) over a period of time, in this case between 1997 and 2012. Flows into the inventory include NPS accessions via the rated pipeline; recalls of personnel separated from any of the three components; and affiliations from one of the reserve components. Flows out of the inventory include military separations; transitions to nonpilot duty, including pilots promoted to O-6 and subsequently no longer considered part of the pilot inventory; and affiliations to one of the reserve components. The objective is for the total inventory to meet the total requirement. For other career fields, flows into and out of the inventory are similar, although the flows between components could look quite different, as pilot affiliations into the reserve component are typically higher than in other career fields.

The Air Force has given top priority to improving its ability to manage its personnel as a total force. But given the complexity of military personnel flows and the variations among career fields, Air Force personnel managers need tools that will enable them to understand historical trends and to evaluate how well requirements can be met. These tools will need to represent a wide range of factors—from external factors such as changes in economic conditions, to internal factors such as changes in accessions or retention. This report describes work conducted by

RAND Project Air Force (PAF) to develop such a tool, including a description of the model and the data used by the model for these types of analyses.

Figure 1.2. Illustrative Personnel Flows, Active Component Pilots

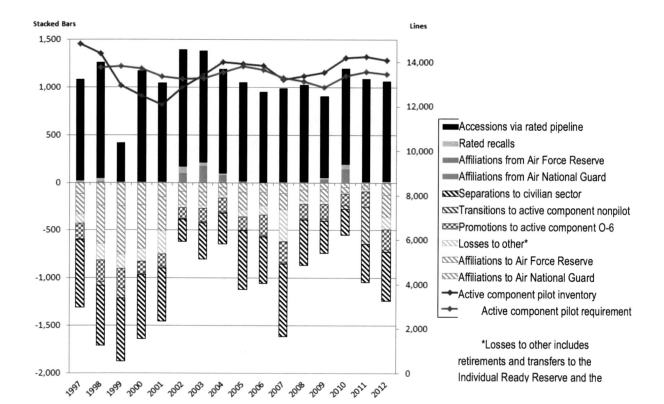

Objectives and Methodology

With a view toward determining what human capital policy changes will be needed to shape the future force size and mix from a total force perspective, the Assistant Secretary of the Air Force for Manpower and Reserve Affairs (SAF/MR) asked RAND PAF to help improve Air Force capability to analyze military personnel flows across the total force, with emphasis on trends and patterns in:

- accessions (by component)
- retention (by component and years of service)
- separations/retirements (by component and years of service)
- affiliations (by component and years of service)
- experience mix across the components (by years of service and grade)
- full-time and part-time manning (for reserve components)
- grade and skill level manning and utilization (by component).

4

The methodology for developing the Total Force Flow (TFF) model is illustrated in Figure 1.3. The first step involved collecting and preparing historical data on personnel, force structure, and economic data. In the second step, we analyze the historical data to identify trends across the total force. Understanding trends and how these trends changed helped to inform our development of model capabilities. Third, we built the TFF with the ability to forecast personnel flows and patterns. We envision that the model will be used to analyze personnel flows—such as the effect on personnel flows of alternative accession or retention rates—and to use this information to develop policy recommendations.

Figure 1.3. Methodology for Developing and Using the Total Force Flow Model

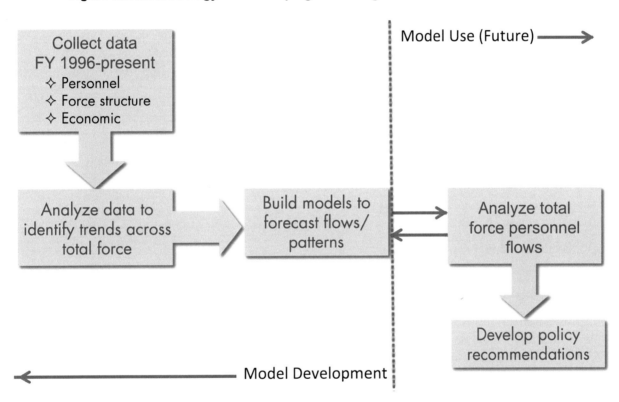

Before discussing the model development in detail, we describe, in Chapter Two, some existing models and tools commonly used to evaluate human capital flows and describe the additional capabilities provided by the TFF model. Chapter Three turns to a discussion of the data collection task—identifying and organizing the data needed by the model—as well as our conclusions regarding the efficacy of available data. Chapter Four describes a new model developed by RAND PAF to extend existing total force modeling capabilities for use in analyzing and forecasting personnel flows. The report concludes, in Chapter Five, with guidelines on using the TFF model and how this capability can be extended in the future.

Chapter Two. Existing Tools for Analyzing Human Capital Flows

A number of analytic models and techniques exist that can be used to analyze personnel flows across the total force. In this chapter we provide an overview of three existing tools: RAND's Total Force Blue Line model, sustainment modeling, and utilization modeling. All of these tools use varying degrees of detail from force structure data to capture the number and composition of Air Force units, to include the major weapon systems inventory and the associated manpower and support equipment authorized based on that inventory.[6]

Total Force Blue Line Model

Traditionally, the inventory of Air Force rated officers has been managed separately in the active and reserve components. For the active component, rated inventory management has been a very intensive process, because it is very costly to train and develop rated officers. In contrast, the Air Force Reserve and the Air National Guard fill most of their rated requirements with officers who previously served in the active component. That supply has been abundant, so comparatively little management attention has been needed.

This model, however, is no longer applicable. Managing the total inventory of rated officers has become increasingly challenging, because the capacity to train and develop new rated officers has declined along with the active component force structure. In addition, a longer active duty service commitment (now ten years for pilots and six years for other rated officers) means that pilots are staying in the active component longer, with fewer available for reserve component affiliation. As a result, it has become increasingly important to coordinate rated inventory management across all three Air Force components.

RAND PAF developed the Total Force Blue Line (TFBL) model to support total force rated management. The tool generates inventory projections—often referred to as blue-lines—for all categories of rated officers in all three Air Force components. These inventory projections are directly tied to

- *production rates* of new rated officers

- rates at which rated officers *separate* from the active component

---

[6] The Air Force analytical community also maintains several models to evaluate AC and RC force structure and cost. The Total Force Enterprise Analytic Framework is a force shaping tool comparing different levels of AC/RC blending at various deploy-to-dwell ratios. The Symbiotic Relationship model uses historical data to analyze the interdependencies of the AC and the RC and analyze how changes in the policy of one component affected the other component. The Air Force Reserve's Individual Cost Assessment Model examines manpower costs associated with AC/RC force mix decisions.

- rates at which separating active component officers subsequently *affiliate* with the guard or reserve.

The model integrates rated management among the three components, which is particularly valuable because it can capture the effects of how personnel flows in one component can impact inventories in other components. The model is able to accommodate representation of a wide range of policies and can be used to assess those policies (such as changes in accession or retention rates). Or, given requirements for specific inventory levels, the tool can aid in determining changes in personnel flows that will most closely approach those requirements. This linkage between personnel flows and inventory for all three components enables the use of this model as a rated management tool for the total force.

The TFBL model estimates rated inventories. These estimates begin with a set of inventory levels for all three components drawn from personnel files provided by the Air Force Personnel Center.[7] The model then estimates inventories for subsequent years using a simple conservation equation: The inventory of any rated category at the end of a fiscal year equals the inventory at the end of the previous year, plus gains during the year minus losses during the year. The model estimates inventory categories that cover each component, crew position, and major weapon system—nearly 60 categories in all—and considers the usual types of gains and losses:

- new officers who complete undergraduate training and earn their wings

- officers who separate from the active component and leave military service

- officers who leave the rated inventory (such as upon promotion to O-6) but who remain in active service

- officers who separate from the active component and affiliate with a reserve component (both a loss and a gain)

- other transfers of officers in and out of the force or between inventory categories.

The model tracks these gains and losses for each component, estimating separation and affiliation rates based upon historical baselines.

Though the model focuses on rated inventories, it also considers requirements. These are the funded authorizations for rated personnel—the jobs that rated officers must perform. The Air Force calls these requirements its rated Red Line. These requirements are inputs into the model and are provided by the Air Force Manpower Planning and Execution System. Like rated inventories, these rated requirements are identified by component, crew position, and requirement type—which, for the most part, corresponds to a major weapon system. By comparing inventory projections to requirements and determining how well available inventories meet the requirements for pilots, it is possible to gauge the health of the rated inventory. The model also has sufficient flexibility to allow the user to assign how inventory categories fill

---

[7] The inventory levels are very similar to the data sets discussed in Chapter Three.

requirements categories. Using "assignment rules" defined by the user, the model will use surplus inventory in one category to fill shortfalls in other categories. The TFBL model also has the capability to allow rated officers from one component to fill inventory requirements in another, should the user choose to employ this option.

Considering alternative force management levers, assignment rules, inventory projections, and requirements, it is possible to evaluate the degree to which the reserve components are able to satisfy their requirements for rated officers by hiring pilots with prior service in the active force—and the corollary, how much of the reserve inventory will need to be filled with pilots who have no prior military service. As a gauge of feasibility, the model's results can be compared to historic affiliation rates and to programmed training capacity.

Figure 2.1 shows how the model can be used to estimate how well the Air Force Reserve and Air National Guard could satisfy their future needs for pilots by hiring pilots separating from the active component. These estimates assumed that pilot accessions with no prior military service would continue indefinitely at currently programmed rates—1,046 total pilots by the active component; 81 total pilots by the reserves; and 141 total pilots by the guard.

How well the guard and reserve can satisfy their future needs for pilots will depend on the

- number of active component pilots who separate
- fraction of separating pilots who affiliate with the reserve component
- rate at which the reserve component loses pilots.

We assumed the same number of separations for all cases. We varied the affiliation rates for each reserve component from 10 to 40 percent of separating active component pilots with 16 or fewer commissioned years of service. We also varied loss rates of guard and reserve pilots. As one would expect, as affiliation rates decline or loss rates rise, the shortfall between inventory and requirements grows. It appears, however, that if the guard and reserve can each affiliate at least 30 percent of separating pilots with 16 or fewer commissioned years of service, they can each come close to filling their requirements. This type of information gives force managers insight as to when problems might arise and the ability to test how changes in personnel flows can mitigate them—changes such as increased affiliation rates, increased capacity, or reduced requirements. An understanding of these effects can serve as a basis for developing effective force management policies.

The TFBL model was designed with the flexibility necessary to allow for modifications to future real-world circumstances. It is one tool that can support management of the Air Force inventory of rated officers as a total force. But the model has limitations. The TFBL model is tailored for analysis of the rated force only and uses historical retirement, separation, and affiliation data to predict future losses without regard to how economic conditions or other external factors might affect these transitions. The model described in Chapter Four extends the capability of the TFBL model to all occupational specialties and incorporates a forecasting component that considers external factors such as civilian economic conditions and pay.

Figure 2.1. Total Force Blue Line Model Estimates of Reserve and Guard Pilot Shortages

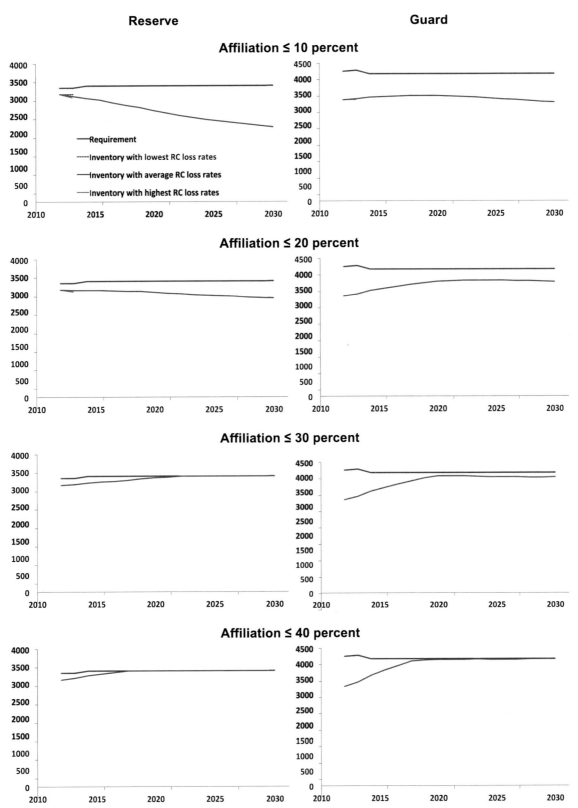

NOTE: Affiliation rates apply to separating pilots with 16 or fewer commissioned years of service. RC = reserve component

Sustainment Modeling

To estimate future needs, workforce sustainment models use historical retention, crossflow behavior,[8] and authorized manpower levels to project personnel-inventory targets for each year of service. These models estimate the flow (accessions, crossflows, separations, retirements, etc.) needed to maintain required personnel levels.

One way to understand personnel force dynamics is to look at historical data for steady-state manning and sustainment. The first step is to calculate continuation rates using data from multiple years for each year of service. A continuation rate is the percentage of personnel in a particular personnel group that continue in service from one year to the next—that is, personnel in year "t" who are in the same personnel group in year "t+1." The product of continuation rates from year of service 1 to year of service "n" is called the cumulative continuation rate (CCR) at year of service "n." Plotting the CCR values from one year of service to 30 years of service yields a full CCR curve over a career. By using multiple years of history (usually five to ten years) it is possible to get an overall view of retention and produce a "steady-state" CCR. With a 100-percent point added at zero year of service, the sum of the area under the curve is the average amount of time one can expect an individual to stay in service. A key assumption is that anything not explicitly modeled to affect continuation rates is the same in the forecast period as it was historically.

We illustrate this procedure with a series of graphs for the career enlisted aviator (CEA) 1AXXX specialty within the Air Force active component. This procedure can also apply to the nonprior service portion of the reserve components, but since much of their inventory is drawn from the active component (denoted as prior service), we would not expect the full inventory to display the same kind of strictly decreasing year-of-service profile displayed by the active component force.

Figure 2.2 shows a CCR for the CEA 1AXXX career group using continuation rates for the period fiscal year (FY) 2003 to 2012. The figure shows a typical decline in inventory to about 60 percent at around the second term of enlistment, followed by very small losses until year 20, and then significant attrition once airmen are retirement eligible.[9]

---

[8] Crossflow behavior is people cross-training into other career fields.

[9] Because many airmen leave at exactly 20 years of service, the continuation rate for year 20 is very low, producing the apparent dive from year 19 instead of year 20. If we measured each record on its EAD (entry on active duty) anniversary instead of at the end of each fiscal year, the dive would start at year 20.

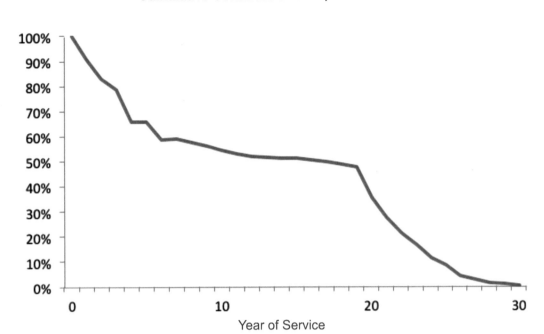

**Figure 2.2. Career Enlisted Aviator
Cumulative Continuation Rate, FY 2003–2012**

Year of Service

Knowing the enduring individual manpower requirements—that is, the permanent party (PP)[10] positions—for the group of interest, it is possible to scale the CCR to contain the total number required under what we call a PP sustainment curve (Figure 2.3 red line). This is done by dividing each CCR value by the expected service longevity—usually called expected man-years—and multiplying by the total authorizations. As a result, the share of total PP positions allocated to each year of service cohort is proportional to the size of the cohort as a share of the total CCR curve.

It must be noted, however, that a significant share of inventory at any point in time is unavailable to fill personnel requirements. This share is called students, transients, and personnel holdees (STP).[11] Historically, STP constitutes about 9 percent of the enlisted inventory and about 15 percent of the officer inventory. Since STP personnel are necessary to sustain the PP portion of the inventory, we must account for them to construct a total steady-state inventory we will call the "objective force." For example, to get the enlisted objective force, we would divide the PP authorizations by 0.91 (that is, 100 percent minus 9 percent STP) to get the full inventory requirement (Figure 2.3, green line).[12] Using this inflated requirement instead of the PP

---

[10] Permanent party authorizations are enduring individual positions.

[11] The personnel holdees can be further categorized as patients and prisoners. Sometimes pipeline students (those in initial training) are pulled out into a separate category, but for our purposes it is sufficient to consider them part of STP. Because those students constitute the majority of STP, the STP account is disproportionately high in years of service 0–2.

[12] By proportionally inflating the CCR curve into an inventory sustainment line, we do not intend to suggest that STP is shared equally across years of service in reality. However, as the CCR was calculated from inventory, not

authorizations gives us another key feature: The value of the objective force at zero years of service is the number of accessions needed to sustain the force at the current authorized strength.

**Figure 2.3. Career Enlisted Aviator
Sustainment Curves, FY 2003–2012**

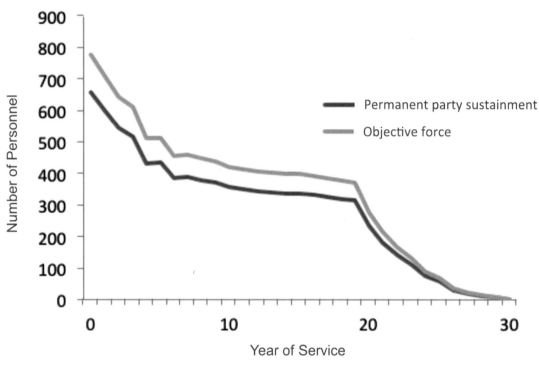

Unit manpower documents (UMDs) designate the grade and, for enlisted, the skill level of each authorization. The manpower community vets the grade and skill distributions at an aggregate level, but this vetting is done at most once a year and assumes very high promotion and retention rates. As a result, UMD requirements across the Air Force and within Air Force Specialty Codes (AFSCs) generally authorize more high-grade and high-skill billets than can be sustained by the authorized number of lower-grade and lower-skill billets, under current promotion and retention trends. Figure 2.4 depicts authorizations, steady-state assigned PP, and steady-state inventory by skill level, for the CEA 1AXXX career group for the period FY 2007 to FY 2012.

We translate personnel by years of service (as shown in Figure 2.3) into skill level by looking at the historical skill levels for each year of service. Each year of service historically has some composition of 3, 5, 7, 9, and 0-level airmen (or Chief Enlisted Manager, CEM). We multiply that observed mix by the corresponding year of service on the inventory sustainment line and then sum by skill across all years of service. Essentially, we are removing the natural peaks and

permanent party, it is appropriate to build an STP factor into the requirement for each year of service and expect the objective force shape to follow the retention patterns already observed.

valleys in the inventory profile and deriving the "objective force" total mix of skill levels. In Figure 2.4, the blue and red bars are directly comparable, but the green bar, which is almost double the 3-level red bar, is not comparable because about half of the 1 and 3-level inventory are STP.

Figure 2.4. Career Enlisted Aviator Inventory by Skill Level, FY 2007–2012

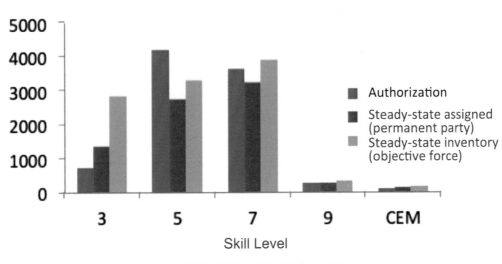

NOTE: CEM = Chief Enlisted Manager

Figure 2.5 shows by skill level the CEA 1AXXX career group manning we would need to fill requirements, given FY 2007 to FY 2012 authorizations and current attrition patterns. The number of 3-levels authorized is far below the number needed to produce the 5-levels the Air Force claims are needed, in large part because the Air Force consistently authorizes more than 91 percent of its congressionally authorized enlisted end strength as PP. Not taking into account the number of airmen likely to be in STP leads to a perpetual state of undermanning at the aggregate level in 5- and 7-level PP billets.

To illustrate the imbalance of authorized skill levels, we can first look at the sustainment curve for the aggregate requirement across all skill levels and reshape the sustainment curve to match authorizations by skill level. Continuing with the CEA personnel group, Figure 2.6 shows inventory by skill level scaled to the total requirement. This figure shows a natural steady-state inventory, eliminating the effect of varying accession cohort size from year to year. Figure 2.7 then uses the same distribution of years of service for each skill level, but scales the size of each skill distribution to the total requirement at that skill level. In other words, the second figure shows the retention levels needed to achieve 100-percent manning at every skill level, for each year of service. The result is a sustainment curve impossible to attain without a significant amount of 5-level cross-training. Nearly every AFSC looks like this, and the aggregate enlisted force does as well. Counting on so much cross-training is not feasible.

**Figure 2.5. Required Career Enlisted Aviator
Steady-State Permanent Party Manning**

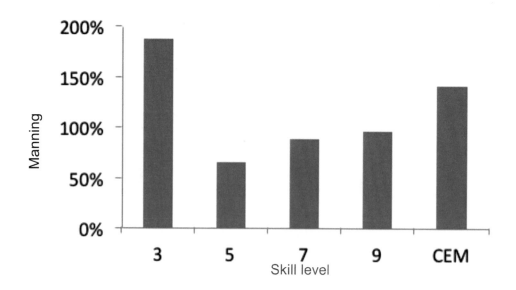

NOTE: Manning by skill level is based on FY 2007–2012 authorizations
and is calculated as the ratio of assigned to authorized personnel. CEM = Chief Enlisted Manager.

**Figure 2.6 Career Enlisted Aviator Sustainment
Using the Aggregate Requirement**

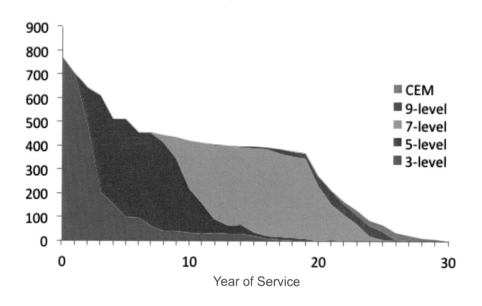

NOTE: CEM = Chief Enlisted Manager.

14

**Figure 2.7. Career Enlisted Aviator Sustainment
Using Requirements by Skill Level**

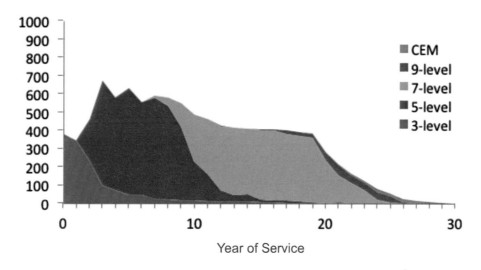

NOTE: CEM = Chief Enlisted Manager.

Using sustainment-modeling techniques, it is possible to identify cases when the manpower community does not enforce sustainable authorization ratios across a specialty, resulting in unrealistic steady-state manning and retention expectations. While sustainment modeling uses historic retention rates to estimate future workforce needs, it does not consider the impact of economic conditions, nor can it adjust for changing manpower authorizations.

Utilization Analyses

RAND PAF's study of total force pilot utilization plowed new ground, as it establishes common metrics for comparing how the three Air Force components use their manpower within a functional group. This type of analysis can reveal alternative ways to optimize the use of human capital across the total force.

As shown in Table 2.1, we used five-character AFSCs (four digits plus the suffix) to break out pilots in the current force as operational, serving in staffs at or below the wing level, serving in staffs above the wing level, or at a replacement training unit.

Table 2.1 Pilot Utilization, Air Force Specialty Code Designation

Operational		Staff at/Below Wing Level		Staff Above Wing Level	Replacement Training Unit
4th digit	Suffix	4th digit	Suffix	4th digit	4th digit
2 or 3	Matches a specific aircraft designation	3	Does not match a specific aircraft designation, such as "general" or "other"	4	1

Figures 2.8 and 2.9 show the number of operational and staff and air liaison positions filled by active, reserve, and guard pilots over time. Evident in the figures is the significant decrease in the number of staff and air liaison positions being filled by active pilots (Figure 2.9), while the number of operational positions remains relatively steady at an average inventory of 8,000 (Figure 2.8). The reserve, however, has increased its staff positions filled from a low of 258 in FY 1997 to 539 in FY 2012.

Figure 2.8. Pilot Operational Positions Filled (calendar year 1996–FY 2012)

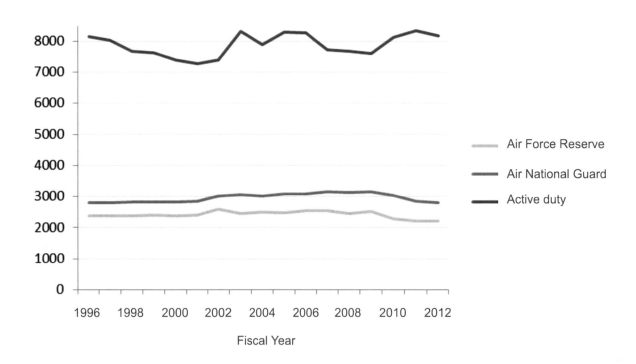

Figure 2.9. Pilot Staff and Air Liaison Positions Filled (FY 1996–FY 2012)

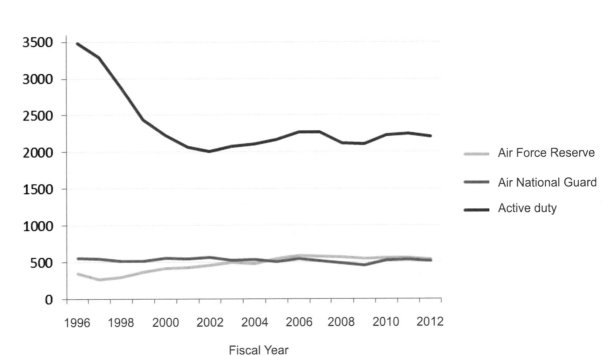

For an indication of the required number of operational positions, we can look at the total number of pilots per aircraft over time. Figure 2.10 shows the pilots per aircraft for each component.[13]

---

[13] We show the number of pilots divided by current fleet numbers for a longitudinal comparison. More detailed analysis was performed using stratification by aircraft mission design series.

Figure 2.10. Pilots per Aircraft by Fiscal Year

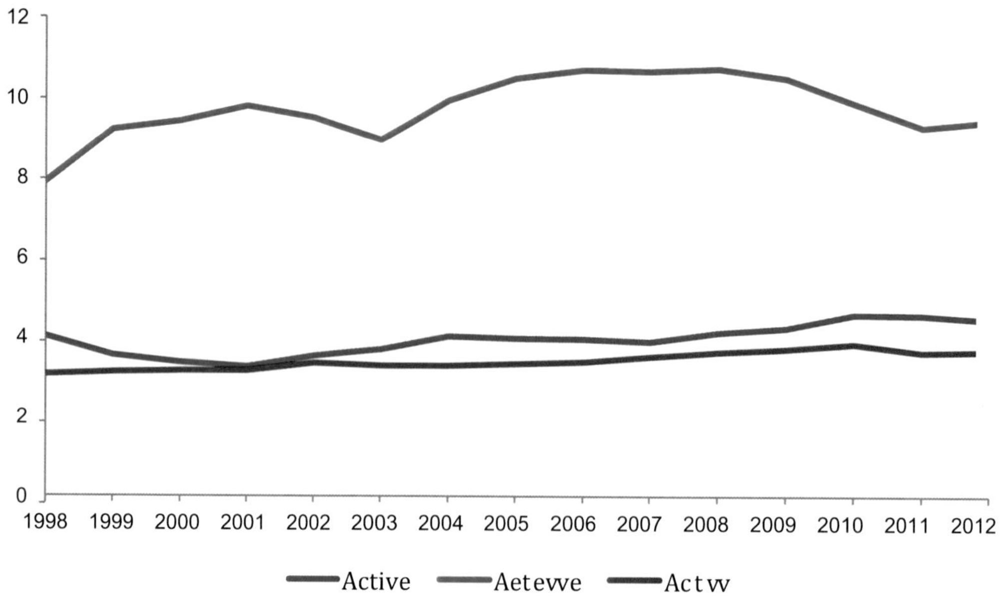

Using Primary Aircraft Authorization (PAA)[14] numbers provided by Air Combat Command for the Combat Air Forces and by Headquarters Air Force for other aircraft, RAND developed yearly force structure data sets in a manner that facilitated integration with the manpower and personnel data sets discussed in Chapter Three. The results demonstrate that pilots per PAA increased for all three components over the study period. Table 2.2 breaks this down further.

---

[14] PAA is the number of aircraft authorized to a unit for performance of its operational mission. PAA serves as the basis for allocating operating resources such as manpower, support equipment, and flying hours. It excludes backup, attrition, and reconstitution reserve aircraft.

Table 2.2. Changes in Pilots per Primary Aircraft Authorization

Component	Pilot Delta (%) 1998–2012	PAA Delta (%) 1998–2012	Pilot per PAA 1998	Pilot per PAA 2012
Air Force Active Component	–2.1	–11.2	4.0	4.4
Air Force Reserve	2.2	–14.3	7.9	9.4
Air National Guard	–2.7	–18.3	3.1	3.6

The total number of Air Force Reserve pilots slightly increased from FY 1998 to 2012. The total number of active component and Air National Guard pilots slightly decreased over the same period. The number of PAA decreased by more than 10 percent for all three components over this same time. Thus, the pilots per PAA increased for all three components.

Increases in pilots per PAA were due to a number of factors:

- reducing force structure and operational pilot numbers by decreasing squadron PAA counts (e.g., downsizing fighter squadrons from 24 PAA to 18 PAA) without proportionately decreasing the numbers of squadrons. This increases pilots per PAA because the overhead tax for group and wing staff positions does not decrease, unless the numbers of groups and wings also decrease

- reducing force structure and operational pilot numbers without proportionately reducing the number of staff organizations and positions above the wing level (numbered air force, major command, component, and headquarters)

- increasing crew ratios

- increases in staff positions requiring rated officers.

Decreasing the numbers of squadrons, groups, and wings commensurate with force structure decreases would have allowed the Air Force to maintain constant numbers of pilots per PAA numbers, but this might have made it more difficult to ensure that all required future senior rated officers had the opportunity to command at those levels. Future utilization modeling must take second- and third-order effects like this into account.

Moving Beyond Current Capabilities

This chapter's discussion of the TFBL model and sustainment and utilization methods shows how each of these tools has contributed to an improved picture of personnel flows across the total force. TFBL can give force managers insight about the timing of cross-component personnel management problems that might arise, along with the ability to test how changes in personnel flows (accessions, retention, etc.) might mitigate them. Sustainment tools help force managers understand career field sustainment demands. Our CEA example demonstrates that when the manpower community does not enforce sustainable authorization ratios across a specialty, the result is unrealistic steady-state manning and retention expectations. Finally, our examination of total force pilot utilization established common metrics for the use of manpower within a functional area across the three Air Force components.

All of the total force management capabilities depicted here are clearly useful for force management. However, as inventory management becomes more challenging, the importance of coordinating across all three Air Force components will continue to grow. Thus, future personnel managers will need additional capabilities to manage the total force. These include the ability to

- forecast future inventories, considering the effects of changes in economic conditions
- examine the implications across the total force of changes in personnel flows that might result from policy changes in a particular component
- examine the interrelated effects of changes to accessions, affiliations, separations, and retirements
- assess actions necessary to more efficiently meet component contributions to the overall requirement.

RAND has developed a new component flow model that adds these capabilities to the suite of existing tools. The next chapter describes the data used in the model, with model specifications and capabilities described in later chapters and appendices.

Chapter Three. Identifying and Organizing Historical Data for Analysis

The component flow model described in the following chapter uses career history data of military personnel and manpower requirements in the Air Force active and reserve components to gain an understanding of personnel flows within and among the components. The data are used in our model to analyze and forecast how a wide range of factors affect these flows, from external factors such as changes in economic conditions to internal factors such as changes in retention and accession rates.

These analyses require an extensive array of data drawn from numerous sources inside and outside the Air Force. A critical factor in retrieving data from these sources is the reliability of the data—data elements and values for data elements must be robust over time. The study focused on data from fiscal years 1996 to 2012, because that represented a time frame during which consistently reliable personnel data are available, but the methodology described in this report is useful for forecasting personnel flows beyond 2012 and is not limited to analysis during this period.[15]

We collected data of the following types:

- *Personnel*. The people serving in Air Force officer and enlisted specialties.

- *Manpower requirements*. The jobs (also called positions or billets) that must be filled to perform Air Force missions.

- *Economic*. Economic growth indicators and employment statistics at the national level and for select industries; occupational wages and employment costs; and postsecondary education enrollment statistics.

This chapter describes the data elements collected, the sources of these data, why they are needed, how they are used, the efficacy (accuracy and utility) of the data, issues that arose in creating the requisite data sets, and how those issues were resolved. Information is provided at the conceptual level; complete documentation of our work is contained in the Statistical Analysis Software (SAS) programs developed. The intent in this chapter is to provide sufficient information so that others could create comparable data sets to employ the model for different time frames, different occupational categories, and other military services.

---

[15] The data can be updated when snapshots at the end of each fiscal year are available.

Personnel Data

Understanding the nature of an Airman's career is critical to our study. The career history documents the behaviors of Airmen in their accession component and throughout their career, during which they may affiliate with another component or leave military service to return to civilian life. Building the career histories required data elements such as component, full- or part-time status, grade, years of service, specialty, date of entry, source of commission or enlistment, separations from military service, and transfers from one component to another. Collecting data of this type and using it to create career profiles at the individual level for personnel across all three Air Force components has never been done before in such a comprehensive manner. The challenge is not only a matter of pulling data from existing personnel files but also of processing the data in a way that enhances its utility in conducting component flow analyses. The uniqueness of the career history data sets produced and of the methodology used to create them makes this study a valuable contribution to future total force personnel management for all the military services.

Personnel Data Elements Extracted

Annual personnel data were obtained from the Air Force Personnel Center (AFPC) at the individual level of detail. AFPC retrieves personnel data from the Military Personnel Data System at the end of each calendar month and manipulates the data in two important ways. First, AFPC assigns a unique identifier code to each individual, thus protecting sensitive name and Social Security Number attributes. We used this unique identifier code to link data records across fiscal year files to create career histories for each individual. Second, AFPC applies numerous data encoding and standardizations producing data fields supporting analysis of personnel data over many fiscal years. AFPC maintains six personnel files we chose to use in our research: active component officers, active component enlisted, Air Force Reserve officers, Air Force Reserve enlisted, Air National Guard officers, and Air National Guard enlisted. This structure provides the flexibility to separately analyze each component's officers and enlisted personnel, consistent with how the components' personnel policies are often implemented.

We created career histories using the following data elements. We use end of fiscal year data, as applicable. In relevant cases, we describe the necessary data transformations.

Component. Individuals are assigned to the active component, Air Force Reserve, or Air National Guard. While this is a relatively straightforward task—given that the military personnel data extract files are generally organized by component—several issues do arise in populating the files. First, some records in the active component files are in fact Air Force Reserve or Air National Guard personnel on full-time active duty. Such individuals have not changed component and must, for our analysis, be associated with their actual reserve component membership. Sorting this out proved difficult, as the components use different data fields and encodings to identify these individuals.

The second issue involves selecting the records of interest from the Air Force Reserve officer and enlisted files. Our analysis includes those members in the Selected Reserve—reserve members who are readily available for call-up to active duty. Other reserve statuses—Individual Ready Reserve, Standby Reserve, and Retired Reserve—are excluded from our data set. The final issue is the occasion when active component personnel appear to be in the reserve component. Prior to FY 2005, new officers not commissioned through a service academy or as a distinguished graduate from another source were given reserve commissions and thus assigned to the Air Force Reserve. These records are assigned to the active component in our data sets.

Career field. For rated officers in the active component, career field classification is derived from the Rated Distribution and Training Management (RDTM) category and the AFSC; for all others, career field is derived from the AFSC.

- *Active component rated officers.*[16] To identify rated officers, we start by finding officers with aeronautical ratings and then include only officers in grades O-1 through O-5, since officers in grade O-6 are not classified as rated. For active component officers with aeronautical ratings in grades O-1 through O-5, we identify the RDTM category for the specific aircraft to which the officer has been assigned.[17] We then use Table 6.2 in AFI 11-412, *Aircrew Management*, to work backwards from the RDTM to the officer's function within the rated officer category: pilot, combat systems officer, air battle manager, or remotely piloted aircraft pilot. If no RDTM code is present in the personnel file, we then use the member's AFSC as we do for both reserve component rated officers and all nonrated officers.

- *Reserve component rated officers and nonrated officers (active and reserve)*. Reserve component rated officers' personnel records do not contain RDTM categories. After identifying aeronautical ratings and officers in grades O-1 through O-5, we categorize reserve component rated officers into career fields using only the AFSC data elements, just as we do for all nonrated officers.

For active component officers without aeronautical ratings, when available, we use the primary AFSC (the AFSC in which the member is most highly qualified) to classify their career field. If the primary AFSC is missing, we use the secondary AFSC in the officer's record followed by the tertiary AFSC, if necessary. If none of these codes is available, we do not include the officer in the career field being analyzed. For officers in the Air Force Reserve or Air National Guard, we use the primary AFSC. If the primary is missing, we use the duty AFSC (the AFSC for the duty the member is performing at the time). If both are missing, we do not include the officer in the career field under analysis.[18]

---

[16] Officers in Air Force aeronautical occupations, or ratings, (pilot, combat systems officer, navigator, air battle manager, remotely piloted aircraft pilot) are commonly referred to as rated officers.

[17] Active component aircrew managers designate their rated officers with a Rated Category for the group of aircraft to which they will be assigned, as well as an RDTM category for the specific aircraft to which they will be assigned. Neither of these designations is used for reserve component officers.

[18] In fact, if we used only primary AFSCs to assign officers to categories, the outcome would be nearly the same, since that data element is rarely missing.

- *Enlisted personnel.* For enlisted personnel in any component, we assign personnel categories using the control AFSC (CAFSC).

Grade. Officer grades are O-1 through O-10; enlisted grades are E-1 through E-9.

Employment category. Employment categories are types of full-time and part-time service. Full-time personnel include all personnel assigned to the active component, as well as a subset of the personnel assigned to the Air Force Reserve and Air National Guard. The remaining reserve component members are part-time personnel. In general, part-time personnel are drilling reserve and guard members serving approximately 39 days per year. Full-time reserve and guard members provide support to the reserve component by "organizing, administering, recruiting, instructing, or training the reserve components."[19] In addition, reserve and guard members are full time when temporarily on orders funded by the active component.

Tables 3.1, 3.2, and 3.3 show how we used data elements available in the various data sets to determine the employment category for individuals across their career histories. As noted previously, personnel data used in our analysis come from different files for active, guard, and reserve. The scheme presented in Table 3.1 is for use with the *active and reserve enlisted* data files. Each of these files may contain members from the active, guard, and reserve component, so the first step is to identify the service component for individuals represented in the file. Once the component is determined, two other data elements in the files (functional category and employment category) are used to determine the employment category and status for that individual. Tables 3.2 and 3.3 show a similar designation for *active and reserve officer* files and for *National Guard enlisted and officer* data files, respectively.

---

[19] The restrictions on duties of full-time reserve and guard members are explicit. Per 32 USC 709 (for the Air National Guard) and 10 USC 10216 (for the Air Force Reserve), the primary duties of air reserve technicians must involve organizing, administering, instructing, or training of reserve component personnel or maintenance of reserve component equipment. They may provide support to federal operations or missions only if it does not interfere with these primary duties. Use of active guard and reserve is similarly restricted under 10 USC 12310 and 10 USC 101(d)(6)(A).

**Table 3.1. Designation of Employment Category and Employment Status
in Active Data Files**

Source Data Element		Designation in Model	
Service Component	**Functional Category**	**Employment Category Title**	**Employment Status**
Active Duty Air Force		Active-Duty Air Force	Full time
Reserve	Serving on active duty and paid by reserve	Headquarters Active, Guard, and Reserve (AGR)	Full time
Reserve	Reserve forces authorization	Limited Period Recall—Air Force Reserve	Full time
Guard	Serving on active duty and paid by reserve	Statutory Tour	Full time
Guard	Reserve forces authorization	Limited Period Recall—Air National Guard	Full time

Table 3.2. Designation of Employment Category and Employment Status in Reserve Data Files

Source Data Element		Designation in Model	
Air Force Reserve Section ID	**Civilian Air Reserve Technician ID**	**Employment Category Title**	**Employment Status**
Various mission areas	Active Guard and Reserve Officer/Airman Only	Active Guard and Reserve	Full time
Various mission areas	Air Reserve Technician	Air Reserve Technician/Dual-Status Technician	Full time
Various mission areas		Traditional Reservist	Part time
Initial Active Duty Training		Traditional Reservist	Part time
Inactive Duty Training		Individual Mobilization Augmentee	Part time
Training		Active Guard and Reserve	Full time
Individual Ready Reserve		Excluded from Model	
Standby Reserve			
Retired			

Table 3.3. Designation of Employment Category and Employment Status in Guard Data Files

Air National Guard Technician ID	**Air National Guard Active-Duty Status**	**Employment Category Title**	**Employment Status**
Air Technician		Dual Status Technician—Permanent	Full time
Air Technician—Temporary		Dual Status Technician—Temporary	Full time
	Active Duty	Active Guard and Reserve—Permanent	Full time
	Temporary Tour	Active Guard and Reserve—Temporary	Full time
Not employed as Air Technician		Drilling reservist	Part time

Entry date. For officers, we use the date of commissioning to mark the beginning of their career for modeling and tracking purposes. Similarly, for enlisted personnel, we use the enlistment date. For the majority of individuals, the Total Federal Commissioned Service Date (TFCSD) and/or the Pay Date provide the correct information.[20] However, in cases where there is a break in service, these dates do not accurately mark the date for entry to military service, since both dates are adjusted forward for breaks in service. Therefore, for all individuals we note their Pay Date and TFCSD in the data sets and then use the earlier of those dates for the date of entry.[21]

Years of service. Completed years of service for officers are calculated from the TFCSD. This date includes all periods of federally recognized commissioned service, whether active or nonactive duty, and it is adjusted for breaks in service. Completed years of service for enlisted personnel are calculated from the Pay Date, which includes all periods of federally recognized enlisted service, whether active or nonactive duty, also adjusted for breaks in service (discussed in a following section).

Entry category. For officers, the entry category is the source of commissioning—either Reserve Officer Training Corps (ROTC), service academy, Basic Officer Training (for the active component and Air Force Reserve), Academy of Military Science (basic officer training equivalent for Air National Guard), or other (including the Commissioned Officer Training course for direct commission officers). We select the value of source of commissioning from the earliest personnel file in which a member appears. For enlisted personnel, scores on the Armed Forces Qualifying Test (AFQT) determine the entry category.[22] In the modeling discussed in the next chapter, transition probabilities for officers of different commissioning sources and enlisted airmen of different skill levels are calculated separately and allowed to differ in their sensitivity to changes in economic conditions. Using this approach, the model can capture differences in propensity to separate and affiliate that may be due to skill levels of enlisted airmen or commissioning sources of officers.

Breaks in service. After determining an individual's entry date, entry category, and years of service (adjusting for breaks), we treat breaks in service as losses in the year in which the breaks took place, and gains when the individual returns from the force. As an example, consider Table 3.4, which presents data from an individual's career history. Each individual has a unique RAND

---

[20] For a definition of TFSCD, see AFI 36-2604, *Service Dates and Dates of Rank*, October 5, 2012, p. 7. For a definition of Pay Date, see DoD 7000.14-R, *Financial Management Regulation*, April 2013, Volume 7A, pp. 1–4.

[21] The Date Initial Entry Uniformed Service (DIEUS) may appear to be the accurate date for these purposes, since it is a fixed date not adjusted for time lost or breaks in service. However, this date includes enlistment during ROTC programs, active component delayed entry programs, and similar designations that are inactive periods of service and are not appropriate to include for force management purposes.

[22] The AFQT score is a percentile score divided into the following categories: Category I: 93–99, Category II: 65–92, Category IIIA: 50–64, Category IIIB: 31–49, Category IVA: 21–30, Category IVB: 16–20, Category IVC: 10–15, and Category V: 0–9. Category IIIA, II, and 1 are the first, second, and third (or more) standard deviations above the mean, respectively. Category IIIB, IV, and V are the first, second, and third (or more) standard deviations below the mean, respectively. In our data sets, Group 0 designates missing data.

ID, and this individual is number 12345. The data begin in the individual's third year of service, when the ROTC pilot was commissioned and served in the regular Air Force. This individual serves for as a pilot for eight years, and by 2007 has completed ten years of service. In 2008, the individual leaves the Air Force but returns in 2009 and completes an 11th year of service, this time in the Air Force Reserve instead of the active component.

Table 3.4. Breaks in Service Example

RAND ID	Fiscal Year	Years of Service	Group	Component
12345	2000	3	Pilots (ROTC)	Active
12345	2001	4	Pilots (ROTC)	Active
12345	2002	5	Pilots (ROTC)	Active
12345	2003	6	Pilots (ROTC)	Active
12345	2004	7	Pilots (ROTC)	Active
12345	2005	8	Pilots (ROTC)	Active
12345	2006	9	Pilots (ROTC)	Active
12345	2007	10	Pilots (ROTC)	Active
12345	**2008**	***BREAK***	**Pilots (ROTC)**	**NONE**
12345	2009	11	Pilots (ROTC)	Air Force Reserve
12345	2010	12	Pilots (ROTC)	Air Force Reserve
12345	2011	13	Pilots (ROTC)	Air Force Reserve
12345	2012	14	Pilots (ROTC)	Air Force Reserve

The model aggregates across individuals in year of service and component cell, and it treats breaks as total force losses when they take place. From Table 3.4, RAND ID 12345 would add one to active losses (and total force losses) in year 2007 and one to reserve gains (and total force gains) in the year 2009.

Note that currently, gains in the model are exogenous and presumed to be set by Air Force policymakers directly. We have not allowed for gains, including gains from breaks, to change with changing economic conditions, but it is possible to extend the model.

Personnel Data Processing Difficulties

Part of the challenge in building cross-component, individual-level career histories is considerations that must be given to data gathered from original data sources. Key difficulties include:

Tracking individuals through careers with changes in component, employment category (full or part time), and Air Force Specialty Code (AFSC). A significant percentage of total force members will enter the active component assigned to a particular AFSC and will serve their entire active component career in the same AFSC until they retire and return to civilian life. These individuals will be relatively easy to track in each yearly file. However, some careers are not as straightforward. An individual may, for example, begin a military career in one of the reserve components assigned to a particular AFSC, transition to the active component for a period of time, leave the active component and return to civilian life, then return to the reserve component as an experienced traditional reservist in a different AFSC. Key to modeling and forecasting the flows of military personnel across the total force is identifying and tracking these movements during an individual's career.

- *Component changes.* We documented each component change within the career histories, but affiliation rates are based only on the first time an individual transferred from one component to another.

- *Breaks in service.* Individuals may have a break in military service after which they return to the same component or join a different component. These individuals and their breaks in service are readily identifiable because no personnel data exist for them during the period(s) when they are not serving in one of the components.

- *AFSC switches.* Most apparent changes in AFSC reflect assignment movement within the same specialty, such as when a bomber pilot moves from a flying assignment to a staff position. To accommodate these movements, we acknowledged a core specialty AFSC, which changes only when an individual moves to a different functional AFSC and remains there for several years.

Changing AFSC designations over time. Since analysis of personnel flows may focus on a single AFSC or on a group of AFSCs, it is necessary to develop a strategy to capture changes in AFSCs (consolidations or separations) so that manpower authorizations and the individuals who fill them can be appropriately categorized into a career field. Historical knowledge of the timing of changes made to specialty codes in the past is necessary to develop an appropriate mapping. Recent changes in the cyber career field serve as an example. In 2010, 33SX[23] Communications and Information officers and authorizations were converted wholesale to 17DX Cyber Operations. When modeling and analyzing today's officer cyber operations personnel flows, it is therefore necessary to classify pre-2010 33SX officers and authorizations as today's 17DXs.[24]

Inconsistency in personnel or manpower data elements across components and over time. Strategies to deal with these data issues often require detailed knowledge of past personnel policies and of the underlying intent of changes to Air Force personnel data systems. This historical research and detailed data manipulation can be a time consuming but necessary process.

---

[23] AFSC codes consist of more positions than we chose to study. The "X" in 33SX, for example, indicates that we did not process the fourth position of that AFSC's value in our data sets.

[24] Before performing analysis based on AFSCs, we translated historical AFSCs to current AFSCs. This normalization is carried out using a SAS program that originated at AFPC. RAND PAF and AFPC jointly developed the program to ensure it meets AFPC and RAND data processing requirements. The officer SAS format is readily applicable to the historical data we are processing. The enlisted SAS format was designed to be applied at the full AFSC level of detail, but this study uses only three-character AFSCs. A revision to the format was made to accommodate our work.

- *Missing data.* Across the yearly files, there is the potential for missing data elements or missing entries in these data elements. This sometimes requires inferring the missing data from other data elements or from data present in earlier files.

- *Changes in the valid range of values over time.* As the personnel system evolves over time and new data elements are added to support policy changes, the allowable entries in a data field may change.

- *Policy changes.* Over time, changes in policies may result in changes in the meaning of the particular value of a data element.

- *Component differences.* Each of the components may have different acceptable entries for data elements and/or may use data elements to indicate different things.

Manpower Requirements Data

Manpower requirements are jobs, often called positions or billets, to which personnel are assigned to perform Air Force missions. These data are documented in UMDs with data elements that identify the skill, grade, AFSC, and other characteristics necessary to perform the duties of the position in a given unit or organization. UMDs are maintained and stored in the Air Force Manpower Programming and Execution System (MPES).[25]

We used extracts from MPES to create data sets that represent the yearly officer and enlisted requirements and contain the following data elements:[26]

- **Component.** Positions are assigned to the active component, Air Force Reserve, or Air National Guard.

- **AFSC authorized.** Positions have an authorized AFSC.

- **Grade authorized.** Positions have an authorized grade. Officer grades are O-1 through O-10; enlisted grades are E-1 through E-9.

- **Enlisted skill level.** Enlisted positions specify one of the following authorized skill levels: 3 - Apprentice; 5 - Journeyman; 7 - Craftsman; 9 - Superintendent; 0 - Chief Enlisted Manager (CEM).

- **Permanent party status.** Position information includes whether or not the person needed to fill the job will be permanently assigned to the unit.

---

[25] MPES is the online management information system designed to collect and disseminate total force execution of programmed end-strength (AFPD 38-2, 5 February 2013). We are assuming manpower requirements stated in MPES are valid and therefore represent the true need. Other RAND research has highlighted the problems with active and reserve component methods for determining garrison and wartime requirements, and the potential for significantly overstated manpower requirements, especially in support functions. Investigations into the validity of stated manpower requirements are needed. See Albert A. Robbert, Lisa M. Harrington, Tara L. Terry, and Hugh G. Massey, *Air Force Manpower Requirements and Component Mix: A Focus On Agile Combat Support*, Santa Monica, Calif.: RAND Corporation, RR-617-AF, 2014.

[26] Like the personnel data, the initial manpower data sets were collected for years 1996–2012, but the model is not limited to this time frame.

- **Full-time or part-time status.** All active component positions are full time. Air Force Reserve and Air National Guard UMDs designate each position either full time or part time.

- **Fiscal years authorized.** Position information is the number of fiscal years in which the particular position has been authorized and the number of fiscal years the position is subsequently authorized after the current fiscal year (for a maximum of five years).

- **Fiscal years funded.** The fiscal years the particular position is funded through the current year and the subsequent five.

These data elements allow for a reasonable estimate of yearly manpower requirements for individual AFSCs, groupings of AFSCs, or other personnel categories.

One area requiring special consideration in developing manpower data sets is how to distribute senior-level positions, for officers and enlisted personnel, across AFSCs. For example, group commander positions for officers have distinct AFSCs[27] but are filled by officers from many different specialties. When we analyze a particular AFSC, these commanders need to be redistributed back into their original specialty to create a complete personnel picture; thus we needed to determine the historical mix of individuals' skills serving in group commander AFSCs. To accomplish this, we sampled the core specialties of past commanders by component and created three-year averages to distribute these personnel.

Economic Data

Changes in the U.S. economy, such as demand for workers and wages, affect Air Force personnel flows, including accessions, retention, affiliations, and separations. Although economic conditions are outside of Air Force control, force planners should understand the influence of changing economic conditions on the behavior of military personnel. To test various hypotheses with regard to that influence, we constructed data sets to measure the strength of the business cycle. We also collected data from forecasts of future economic performance so that we could evaluate how future expected economic conditions will affect future Air Force total force personnel flows.

Wages by career area. The U.S. Census Bureau conducts a monthly survey of households for the Bureau of Labor Statistics (the Current Population Survey). This survey provides a comprehensive body of data on the labor force, employment, unemployment, persons not in the labor force, hours of work, earnings, and other demographic and labor force characteristics. We extracted wages for the relevant career groups for the fiscal years of interest. These data can be used to estimate earnings for several experience profiles, to match employment opportunities with manpower/personnel categories.

---

[27] A 10CO is an operations group commander, 17CO is a cyber group commander, 20CO is a logistics group commander, and 30CO is a mission support group commander.

Real gross domestic product (national). Real gross domestic product (GDP) is an inflation-adjusted macroeconomic measure that reflects the value of all goods and services produced in a given year, expressed in base-year prices. Unlike nominal GDP, measured at current price levels and currency values, without factoring in inflation, real GDP can account for changes in the price level and provide a better basis for comparison when tracking economic output over a period of time. GDP data were collected from the Federal Reserve Economic Data (FRED), Federal Reserve Bank of St. Louis.

Unemployment rate (national). The unemployment rate is estimated by the U.S. Bureau of Labor Statistics to represent the number of unemployed as a percentage of the labor force. The labor force comprises people 16 years of age and older who are either working or actively seeking work, and who currently reside in one of the 50 states or the District of Columbia, do not reside in institutions (e.g., penal and mental facilities, homes for the aged), and are not on active duty in the Armed Forces.

Summary

Human capital management efforts benefit from modeling tools that can be used to assess potential requirement shortfalls, needed policy adjustments, or more efficient ways to structure the workforce. These models have, at their foundation, historical data on personnel behavior, an accurate assessable statement of the manpower required, and accurate representations of key external factors such as economic conditions or civilian wages. We have examined the data sets described in this chapter and, with the data filtering methods outlined previously, judge that these data are more than sufficient for effective human capital modeling efforts. A newly developed model using these data sets is discussed in Chapter Four.

Chapter Four. A New Model for Assessing Human Capital Flows Across the Total Force

Given the complexity of military personnel flows across the service's three components, combined with the variations among career fields, Air Force personnel managers need tools that will enable them to understand historical trends and to evaluate how well requirements can be met from a total force perspective. RAND has developed a new model with these capabilities called the Total Force Flow (TFF) model. As Figure 4.1 illustrates, the model uses the personnel career histories and historical manpower data described in Chapter Three for two primary purposes. The first is to plot and analyze historic data trends. The second is to forecast personnel flows and to evaluate and optimize the ability to match inventory to requirements based on different assumptions about manpower requirements—such as new end strength targets, results of personnel readiness evaluations, proposed changes to organization and mission, and adjustments to the mix of active and reserve forces.

The remainder of this chapter describes the capabilities of the forecasting and optimization components of the model.

Figure 4.1 Total Force Flow Model

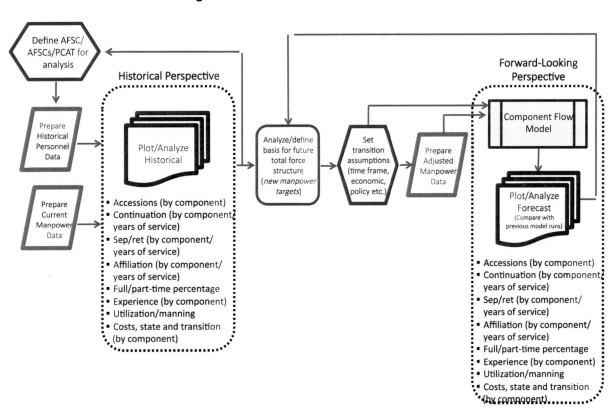

Forecasting

A major simplifying assumption implicit in many existing manpower models is that important parameters are fixed and unresponsive to external forces, such as varying economic conditions. This is a major limitation when modeling the Air Force's military personnel flows, because all U.S. military members are volunteers drawn from the civilian labor market, where individuals have many different employment options. Changes in economic conditions can therefore have significant effects on the Air Force's ability to hire and retain its workers.

To help the Air Force manage those effects, we have developed a capability to estimate the historic impact of changes in economic conditions on the flows of labor into, between, and out of the Air Force active, Air Force Reserve, and Air National Guard components. Using those estimates, the model is able to forecast what is expected to happen to military personnel flows based on different scenarios for future economic growth. It is also able to estimate the effect of changes to a particular personnel flow (perhaps from a proposed policy change to accessions, retention, etc.) on total force personnel flows and inventories. Here, we provide a brief description of the forecasting approach. Details on the model are in Appendix A.[28]

In general, the historical portion of our model uses the personnel career histories described in Chapter Three to track individual military members throughout their Air Force careers, from initial accession to final separation, including any transitions they make between components, via direct affiliation or after temporarily separating from and later returning to the military (a "break in service"). In each year, we decompose the total inventory into periodic snapshots like those depicted in Figure 4.2. We call these snapshots "states." Each state is a fiscal year's group of individuals in a common component who share a common entry category, a common employment category, and a common level of experience as defined by years of service. As the figure shows, in any given year, personnel in a given personnel category and with a given level of experience enter a component from the civilian sector or one of the other two components. When personnel exit a given state they either affiliate with another component or return to civilian life. The user of the model can determine the set of requirements and personnel to include in a model run—everyone in the Air Force, a particular AFSC, a group of AFSCs, or any other personnel grouping.

---

[28] Another RAND model, the dynamic retention model, is also capable of similar estimation and forecasting but is more sophisticated and computationally intensive than the model developed here. The dynamic retention model is a tool able to assess the effect of alternative compensation proposals on active and reserve component retention. See Beth J. Asch, James R. Hosek, Michael G. Mattock, and Christina Panis, *Assessing Compensation Reform: Research in Support of the 10th Quadrennial Review of Military Compensation*, Santa Monica, Calif.: RAND Corporation, MG-764-OSD, 2008; and Beth J. Asch, Michael G. Mattock, and James R. Hosek, *A New Tool for Assessing Workforce Management Policies Over Time: Extending the Dynamic Retention Model*, Santa Monica, Calif.: RAND Corporation, RR-113-OSD, 2013.

Figure 4.2. Depiction of Notional States in the TFF Model

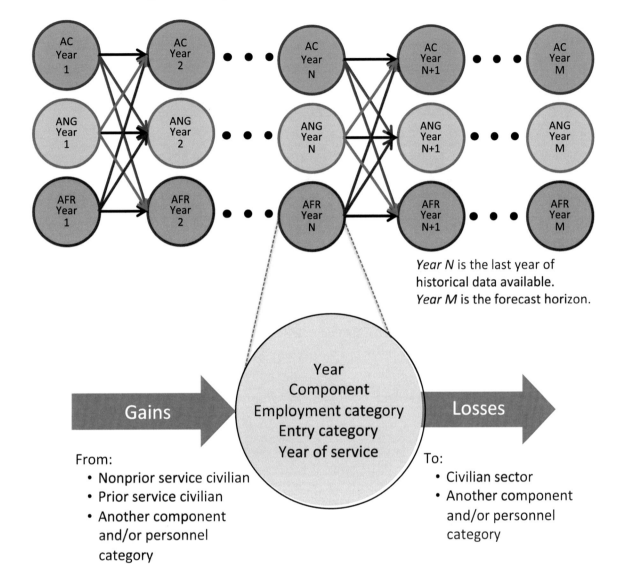

We use FY 1996 as our first year of historical data. Starting there, we follow the career paths of individual members from one fiscal year to the next, tracking their flows within and between components, thus arriving at new states in each subsequent fiscal year. This continues throughout the chosen time frame of available historical data. Based on the historical patterns thus identified, the model calculates what we call "transition probabilities," which reflect the likely flow of service members into and out of each state. These probabilities can then be used to predict future states. The next section explains the process used to estimate the impact of changing economic conditions and illustrates why incorporating such a capability into our modeling provides a valuable tool to force managers.

Transition Probabilities and Changes in Economic Conditions

Military careers tend to follow certain patterns, such as when individuals are most likely to leave active component military service, which often corresponds with the end of an individual's initial active-duty service commitment. Another typical pattern is the exodus from the force that occurs after 20 years of service, when service members are eligible to receive retirement benefits. In general, the likelihood that an individual will transition from the active to the reserve component or leave military service altogether varies across two dimensions: (1) years of service, which capture career paths, and (2) fiscal years, which capture environmental factors.

Using pilots as an example, Figure 4.3 illustrates these characteristics. In both of the fiscal years reported (FY 1999 and FY 2008), there is a very low probability that pilots will transition from the active to the reserve component either early in a pilot's career or late in a pilot's career. However, in mid-career, between eight and 18 years of service, the transition probabilities increase dramatically. Much of the year-of-service variation in transition probabilities is due to unique features of the pilot career field, such as active-duty service commitments and retention bonuses. Thus, calculations of transition probabilities must take the year-of-service dimension into account. Furthermore, the model must also be able to calculate these probabilities separately for different career fields, as career paths differ by career field.

Figure 4.3. Pilot Transition Probabilities from Active to Reserve

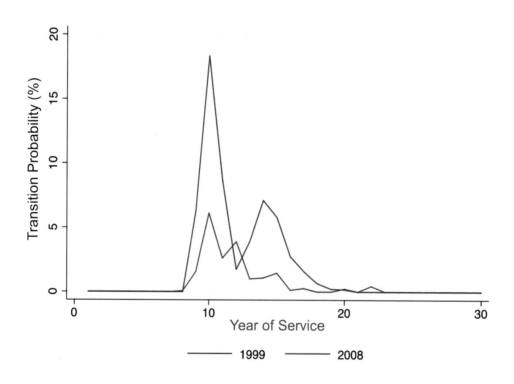

SOURCE: Authors' calculations.

Another important feature of transition probabilities is that they are not constant over time nor across fiscal years. Instead, they are influenced by external factors such as economic conditions, as well as internal factors such as Air Force personnel policies. To illustrate this point, Figure 4.4 shows the real growth in the U.S. GDP between 1990 and 2015. Comparing the trends in GDP to the transition probabilities in Figure 4.3, it is easy to see a relationship between the two.

Figure 4.4. Real Growth in Gross Domestic Product, Actual and Forecasted

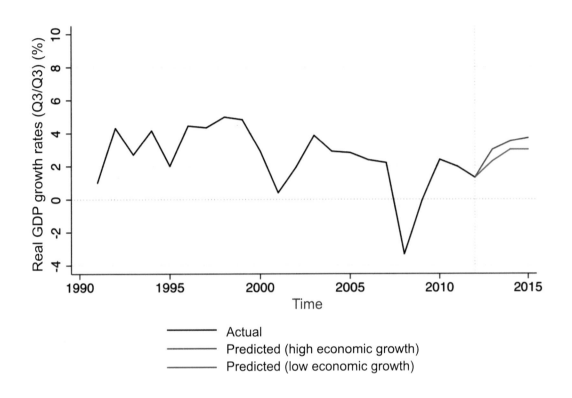

SOURCE: Bureau of Economic Analysis

In 1999, there was more than a 15 percent chance that a pilot would transition from the active component to the reserve component at ten years of service, while the probability in 2008 was only about 6 percent. These differences correlate with very different conditions in the U.S. economy. In 1999, the United States was ending a decade of rapid economic growth at a pace of as much as 5 percent per year, driven by productivity improvements created by innovations in information and computing technologies. In 2008, the United States was in the midst of several years of economic decline, after the collapse of the housing market and the associated financial crisis. In the earlier robust economy, pilots were more willing to transition from the active to the reserve component, where they could take advantage of the greater employment opportunities and higher wages provided by the civilian labor market while also flying for the reserves. When the economy slowed, pilots were less willing to transition out of the active component, increasing retention.[29] It is important to note, however, that the majority of transitions in both cases happened at natural transition points during a pilot's career.

Our model takes this type of relationship into account when estimating transition probabilities. The model evaluates the probability for every possible transition decision a service member can make while remaining in the service (active-to-active, active-to-reserve, active-to-guard, reserve-to-active, reserve-to-reserve, reserve-to-guard, guard-to-active, guard-to-reserve, and guard-to-guard), as well as any decision to leave military service entirely (active-to-separation, reserve-to-separation, guard-to-separation), after taking into account year-of-service effects and fiscal year factors.

Our measures of economic activity include estimates of the median wages that members of each career field could earn working in the civilian labor market, real GDP growth rates, and unemployment rates. Using these measures, we estimate transition behaviors separately for each career field, since the pilot career path can look very different from that of nonrated operations, logistics, and support personnel, and members of each career field may react differently to varying economic conditions, as well. Within each career field, we also take into account the differences in individual behaviors for different officer and enlisted entry categories, and enlisted skill levels.

Once specified, the model's parameters are estimated using panel data regression techniques,[30] which allow us flexibility to account for differences in the transition probabilities for different years of service and to isolate the impact of external and internal factors. After we estimated the relationship between transition probabilities and economic activity, we used available forecasts to project what would have happened to transition probabilities under high economic growth and low economic growth scenarios.[31]

Our forecasts require unbiased estimates of the relationship between economic activity and transition probabilities. If other factors, such as wartime demands for personnel or changes in retention and affiliation policies, are correlated with economic conditions but omitted from our regression model, our parameter estimates might be biased. Bias could occur because we attribute too much (or too little) of the changes in transition probabilities to economic conditions, when some of those changes were actually caused by other factors. In principle, this problem could be partially addressed by collecting more data and adding more variables to the regression models, and future work will endeavor to do this. However, ultimately, unless we have a natural experiment or another identification strategy, there might always be some omitted variable that

---

[29] Additional work would have to be done to correlate the effects of wartime operations in 2008, and the lack thereof in 1999.

[30] Panel data refers to multidimensional data containing observations obtained over multiple time periods for the same entities or individuals.

[31] In macroeconomics, consensus forecasts are predictions of future economic growth created by combining several separate forecasts that have often been created using different methodologies. We develop high and low economic growth scenarios by using the lower and upper range of these forecasts.

we cannot capture. This is a limitation of our approach, and it is why we not only report projections sensitive to changes in economic conditions but also report projections based on historical transition probabilities.

Stocks and Flows[32]

Of course, the transition probabilities generated by the model are only part of the explanation for observed inventory patterns. The number of people in a career field and at a given year of service in any given fiscal year depends on the number of people in that career field at the previous years of service in previous fiscal years, as well as on the transition probabilities. Some career fields may contain mostly junior members with fewer members in mid-career and later. Others may have larger inventories in mid-career and up to the 20-year retirement point. In this regard, the inventory stocks can exhibit significant momentum, and, barring unlikely dramatic shifts, much of what the inventory will look like a few years hence is contained in the shape and structure of the force today.

After we predict transition probabilities on the basis of economic factors and other year-of-service–specific effects, we use a simple inventory adjustment model to predict what the stocks of inventory will look like in the next year of service, based on previous year inventories, the transition probabilities, and projections of gains to the force. Such adjustment models are also an important part of previous modeling work, such as the TFBL model.

Reduced Form Versus Structural Modeling

In modeling natural and economic phenomena, there are two different types of approaches: structural and reduced form estimation. With a *structural modeling* approach, researchers attempt to use data to identify the parameters of an underlying model of behavior. As an example, to model the flow of Air Force officers between components, researchers might first specify an objective function that individuals maximize by choosing an optimal sequence of career options over the course of their lifetime. The benefits of choosing different options may include observed factors, such as wages, living stipends, or bonuses, and unobserved factors, such as an intrinsic motivation for working in a particular career. By making assumptions about the functional forms of the objective function, how individuals solve the model, and the distribution of the unobserved components of the model, researchers are able to estimate the unknown parameters of the model. This is the approach taken in RAND's dynamic retention models (see, for example, Asch, Mattock, and Hosek, 2013).

Structural modeling has several advantages that make it especially useful for certain research questions involving detailed policy analysis. For instance, a dynamic retention model can be used directly to evaluate the impact of changes in wages or retention bonuses. After estimating

---

[32] A *stock* value is measured at one specific time and represents a quantity existing at a point in time. A *flow* variable is measured over an interval of time.

parameters of the model, researchers can use counterfactual values of different variables, plug those into the underlying equations, and see how individual choices would be different. Structural models can also be used to conduct welfare calculations, which show how much better off individuals are from the policy changes that could have been made. This can be helpful in searching for optimal policies.

However, structural models also have disadvantages. They can be difficult computationally and require a great deal of computer time to solve. The welfare and counterfactual policy analysis that they deliver are accurate only insofar as the underlying model of behavior correctly portrays how individuals make decisions. Often, technical assumptions are used to make structural models tractable and solvable, but these assumptions sometimes make the model unrealistic.

On the other hand, *reduced form* models use regression techniques to fit data, using historical correlations and extrapolation to make predictions about what might happen in the future. They can be used to infer the relationship between outcome variables, such as transition probabilities, and independent variables, like changes in economic conditions. For some purposes, such as making short-term predictions under different broad scenarios, a reduced form approach, which requires few assumptions and is easily estimated, is desirable. At its heart, the TFF model is a reduced form model of inventory in the total force. The model allows changes in economic conditions to change transition probabilities, and this information is used in how predictions about the future are made, but it cannot be used to conduct detailed policy analysis.

That said, certain policy actions can be altered by the researcher, in order to provide insight about what might happen. Accessions are not forecasted but are modeled as a choice variable; the researcher can examine how changes to accessions would impact career fields over a short time horizon. Additionally, the transition probabilities themselves can be directly nudged up or down, depending on whether the policymaker is interested in increasing retention or affiliation rates. However, the underlying policy actions that would lead to these changes, such as changes in recruiting strategies or retention bonuses, are not directly modeled.

Optimizing Current Inventory to Future Requirements

Matching inventory to requirements is the basic task of military force managers. In undertaking this task, managers need to consider the size of the personnel pool, personnel experience levels, how long it takes a typical individual to gain adequate experience to fill a requirement, whether candidates have the proper training to fill a requirement, whether candidates are in the right career specialty, from which component to draw personnel, and other similar factors.

In this section, we provide an overview of a new optimization approach that moves an inventory of individuals through time from an initial position toward future requirements. This optimization approach is a complement to the TFBL model described in the previous chapter.

Whereas the TFBL model generates inventory projections based upon previous behavior, this optimization approach determines how baseline affiliation and transition rates must change to align current inventory with future requirements.

There are several benefits to such an approach. First, it can assist in determining whether it is even feasible to move a given inventory toward a given requirement under different policy constraints and varying relevant external influences. Second, it can assist in evaluating the cost involved and how quickly the change can reasonably be accomplished. More importantly, it can assist force managers in identifying policies that balance the time required against the likely cost.

While this optimization approach can be powerful, it was designed for application to the flows of human capital across the entire total force. As with the other tools presented in this report, we have not attempted to model the movement of specific individuals to meet requirements. Other methods are better suited to such detail, such as RAND's dynamic retention model.

Objectives and Constraints

The optimization approach described here moves an inventory of individuals through states in time, as explained above, from an initial position toward a requirement.

During each fiscal year under consideration, the model (1) specifies the *number of individuals accessed into each state from outside the total force* and (2) identifies the *number of individuals who transition from one state to another*—that is, for example, the number of Air Force active component logistics officers whose commissioning source was the U.S. Air Force Academy who transition from 10 years of service in the active component to the Air National Guard with 11 total years of service. In order to identify the feasible amount of time necessary to bring the inventory in line with the requirement, we constrain the model's ability to reshape the inventory by imposing bounds on the number of individuals the Air Force can access and the number who can move from one component to another, bounds based on assumed economic conditions and policy constraints. How the model reshapes the inventory gives force managers visibility into areas where inventory shortfalls or overages may exist and can help those managers identify and evaluate policy changes that may eliminate gaps between inventory and requirements—such as bonuses designed to increase or decrease retention in a certain specialty.

In summary, the model makes decisions about how to rearrange inventory based on three objectives it addresses in sequence, with each level's results influencing the next. The first objective constrains the aggregate number of individuals who must be present in each component over the time period being examined. The second objective moves inventory in an attempt to achieve the required experience distribution in each component, while maintaining the total number of individuals in each component. The third objective attempts to minimize the number of changes to baseline accession and transition rates required to adhere to the first two objectives.

These objective functions were structured with a view toward minimizing the total personnel transitions associated with the required inventory levels. We currently have minimal insight into the cost structure associated with making changes to the baseline accession and transition rates, but the aim is to build this capability into the model in the future, so that the model will evaluate the ability to deliver inventory requirements while both minimizing the number of personnel transitions and minimizing cost. Appendix B contains more detail on how the model functions.

Model Inputs

This new optimization approach requires several important inputs, described in Chapter Three. The first major input into the model is the actual current inventory to be considered, drawn from personnel files provided by the Air Force Personnel Center. The second input is the baseline accession and transition rates. Because the optimization model was designed to examine the time and cost involved to match a current inventory to requirements under different external factors (e.g., economic conditions), the baseline accession and transition rates are obtained using the methodology discussed in the previous section on forecasting. The final input into the optimization component of the model is the initial constraints placed on accession and transition rates. Based on historical trends, the model places limits on the maximum accession rate into each component in a given fiscal year and for a given entry category, and on the maximum number of members who can transition from one component to another. The model user can decide whether the maximum accession and transition rates used to determine the forecast are equal to the maximum observed rates over the years included in the historical data or the average of the rates in the historical data.

Example Outputs

What kind of results does the total force human capital flow model produce? Here, we provide several example outputs that illustrate the model's capabilities.

Figure 4.5 shows the historic and forecasted inventory for the Air National Guard portion of the enlisted security forces career field (corresponding data are available for the active component and for the Air Force Reserve). Historically, the guard has higher inventory as compared to requirements in this career field, often referred to as overmanning. During the forecast period, the aim was to match inventory levels to requirements, with annual personnel flows limited by average accession and retention rates.

Figure 4.5. Air National Guard Enlisted Security Forces, Historic and Forecasted Inventory versus Requirements

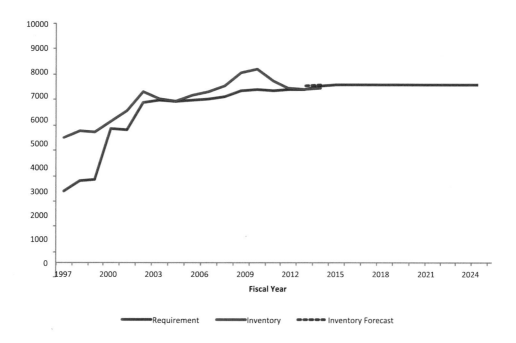

Figure 4.6 shows the FY 2013 inventory by year of service as compared to the model's predicted FY 2020 inventory. The results indicate that it is possible for the guard to regulate gains and losses to the enlisted security forces career field so as to meet the required year-of-service inventory by FY 2020. As with the previous example, personnel flows are limited by average accession and transition rates.

Figure 4.6. Air National Guard Enlisted Security Forces, FY 2013 and FY 2020 Inventory versus FY 2020 Requirements

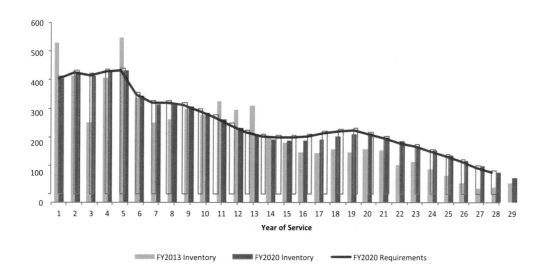

43

The next two figures show the gains and losses necessary to meet the required FY 2020 inventory. Figure 4.7 shows the guard enlisted security forces gains—those flows from the active component (and possibly from the Air Force Reserve), experienced gains (civilian individuals who served in one of the components in the past), and nonprior service accessions. The parameters for this model run constrain the gains to the average of those actually occurring historically. As the figure shows, the necessary gains in each category for the future (approximately 740 total; 80 from other components, 320 experienced gains, 340 nonprior service gains) are reasonable given annual gains attained in the past.

Figure 4.7. Air National Guard Enlisted Security Forces, Historic and Forecasted Gains, FY 1996–2024

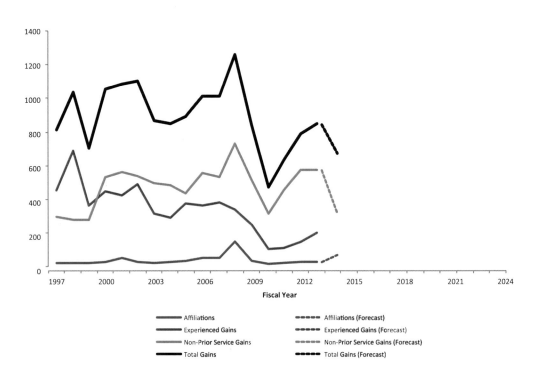

Figure 4.8 shows historic and forecasted losses for the guard enlisted security forces—both separations and retirements and flows to the active and reserve components. Since the parameters for this run of the model are set to meet manpower requirements in future years, losses in FY 2014 and 2015 are less than those in previous years—from an average of 712 between FY 1996 and 2013 to 437 in FY 2015. This decrease in retirements and separations in the guard enlisted security forces career field would need to be accomplished through policy means such as increasing high-year-of-tenure limits or methods to encourage retention. Further runs of the model with adjusted parameters could smooth these losses over several years, and/or help policymakers evaluate other personnel management options.

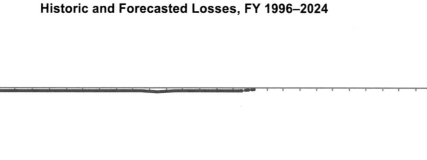

**Figure 4.8. Air National Guard Enlisted Security Forces,
Historic and Forecasted Losses, FY 1996–2024**

The guard's ability to meet future requirements for active-duty enlisted security forces depends on the flows coming from other components and from the civilian sector. The strength of the TFF model is that it provides manpower and personnel managers insight into the effect of personnel policies (such as policies that affect accessions, retention, etc.) on personnel flows across the total force in an integrated manner.

Figure 4.9 illustrates an additional capability of the model: the ability to forecast the effect of changing economic conditions on component flows—in this case, for the pilot force with historical rates from FY 1997–2011 and forecasts for FY 2012–2015. The left-hand side of the chart labeled "Active" shows losses due to separations and retirements from the active component and the number of pilots who leave the active component to affiliate with the reserve or guard. The right-hand portion of the chart shows the gains to the reserve component pilot force—nonprior service pilots with no experience and those experienced pilots who have separated from the active component. In the early years (FY 1997–2001), where active losses were high, we observe a corresponding higher number of reserve component affiliations— especially for the Air Force Reserve.

The spike in active losses in FY 2007 does not result in a significant increase in affiliations, but rather a gradual increase over the years FY 2004 until present. This corresponds to significant increases in civilian airline pilot hiring and improving economic conditions. In the TFF model, we estimate the historic impact of changes in economic conditions on the flows of labor into and between the active and reserve components. Using those estimates, the model is able to forecast what is expected to happen to personnel flows depending on different scenarios for future economic growth.

Figure 4.9. Effect of Changing Economic Conditions on Personnel Flows

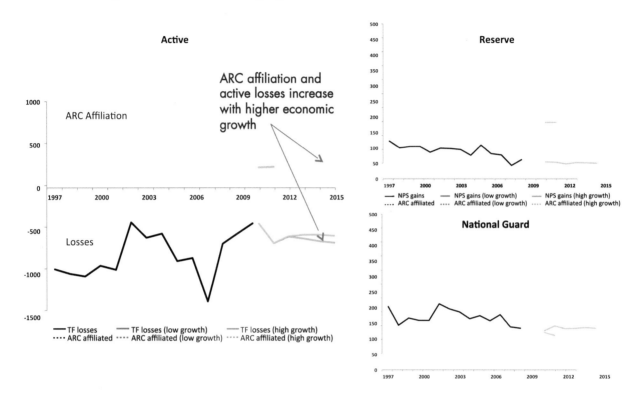

NOTE: ARC = Air Force Reserve Components

The ability to understand the effect of changing economic conditions and to forecast these effects into the future can help personnel managers understand the degree to which economic factors are important with respect to personnel behaviors and consider whether policy changes might be in order to mitigate potential effects. The model's flexibility allows users to consider a range of economic alternatives—in this case, alternative growth scenarios—as depicted in Figure 4.9.

In Chapter Two, we discussed existing methods for modeling sustainment and provided an example using CEAs in the active component. Modeling sustainment for the reserve component is somewhat more complicated than for the active component because CEAs in the reserve component can enter the specialty as crossflows from other AFSCs within the reserve component or as experienced CEAs from the active component, and because this entry to the AFSC can occur across a wide range of years of service. We envision using the TFF model to review historic flows and to forecast combined active and reserve component sustainment to identify trade-offs between active and reserve component authorizations and entries to the career field.

We also discussed, in Chapter Two, methods for analyzing the actual utilization of personnel within a career field using the example of pilots serving in operational versus staff positions. The TFF model allows for the analysis of historic flows and the forecast of future flows for personnel within some subset of authorizations within a particular career field or group of career fields. The only limitation to this type of analysis is that some data element must be available to classify the position as in or out of the set under study.

Future Developments

This chapter has described the baseline TFF model. As with any modeling effort additional capabilities are envisioned for future versions of the model. Perhaps the most important is the addition of cost considerations in three successive steps.

The first cost consideration is to add to the model the average annual costs for each year of the historic and forecast inventories. Given the outputs from the model as it is currently configured, we would calculate the annual cost for the inventories in each state of the model recognizing that there are different costs for each year of service, component, and employment category. In this way, the cost of different personnel flows that might result from policy options could be compared.

The second cost consideration is the inclusion of transition costs. These are costs associated with flows into, out of, and between components, such as the costs for recruiting, training costs for accessions and for experienced gains, and retention bonuses. Such costs are not often considered separately in personnel modeling, but including them enables a more realistic assessment of personnel policies and actions.

A third aspect of integrating cost into the TFF model would include cost as an element of the optimization portion of the model, so that the model could identify the least-cost stocks and flows across the components. This approach would require a more complicated update to the model as compared to the two considerations above. In addition, any cost optimization methodology should understand cost in the context of the overall suitability of model results; that is, the model results need to first offer feasible, sustainable results, and then costs can be considered.

Aside from cost, another potential improvement to the model would be to develop methodologies to deal with small sample sizes. When modeling single AFSCs, the number of individuals in each state may be too small to accurately estimate personnel behavior.

Chapter Five. Final Thoughts

This document describes the TFF model developed by RAND as well as the details for anyone wishing to construct a similar total force flow model for another population. Through the development and testing of the TFF model, we have identified a number of guidelines for how the model can be most effectively used to analyze personnel flows.

First, AFSC-level analysis is often the most appropriate and the most revealing. Analyzing individual AFSCs offers insight into the unique personnel behavior in particular career fields— the level at which personnel is organized and managed. Therefore, analysis results reported at the AFSC level will be most beneficial in determining whether policy changes might be required to meet personnel goals.

Second, having information on historical policy changes enhances the analysis that can be accomplished for a particular AFSC. For example, in several AFSCs we observed wide swings in the yearly number of accessions or significant increases or decreases in the number of separations. Understanding the reason for these changes—and thereby validating them—can greatly contribute to the accuracy of and confidence in the model's forecasts. We know, for example, that Program Budget Directive 720 (PBD-720), released in December 2005 and implemented in FY 2006–2009, planned to reduce over 40,000 manpower positions.[33] This directive had a significant effect on the force management actions for individual AFSCs. Except in special cases such as pilots where manpower targets were being maintained, for most AFSCs annual accessions decreased, retention decreased, and overall inventories decreased to meet the lower targets. Being able to analyze the effect of PBD-720 on past personnel flows increases the confidence we can place in model forecasts should the Air Force face a similar downsizing environment in the future.

PBD-720 was a widely publicized and tracked Air Force-wide force management action. Unfortunately, many policy actions that can have significant effects on personnel flows are not so widely known or tracked. For example, reenlistment bonuses are often used when AFSC personnel managers want to retain enlisted personnel in particular grades or years of service. These bonuses are offered in different amounts, at different reenlistment points, and for different AFSCs depending on force management needs. But historical data on reenlistment bonuses is largely unavailable and, therefore, cannot be used to systematically model their effect on personnel flows. The historical information gathered to support this model would be significantly more valuable if it could be tied more readily to the personnel actions that drove

---

[33] Air Force Audit Agency, "Air Force Personnel Reductions—Audit Report F2008-0004-FD4000," May 12, 2008.

the observed behavior. Looking forward, we recommend that the Air Force improve the way that it documents personnel policy actions (action, population impacted, timing, etc.).

A third consideration when using the TFF model is to approach analyses iteratively, rather than trying to understand influences on predicted personnel behavior all at once. For example, instead of testing the combined impact of a decrease in active component retention and an increase in nonprior service accessions on Air Force Reserve inventory, model each of these changes individually by changing a single parameter and observing the resulting effects across *each* element of the total force. Then, when the individual changes are relatively well understood, the user can model combinations of changes. The value of the TFF model is that it can forecast the multidimensional interactions in active and reserve component personnel flows (accessions, separations, retention, etc.); however, it is also useful in an experimentation setting to develop a deeper understanding of individual interactions as a context for modeling total force flows.

Current TFF model capabilities, while significant, could be expanded, and there is room for further research and model development. In particular, in the current version of the model, the role of uncertainty in future predictions and their effects on optimal choices for policymakers have yet to be fully explored. For instance, consider a case where the Air Force is particularly concerned about meeting certain requirements in the future to ensure readiness, but transition probabilities are forecasted to be uncertain, either because of parameter uncertainty or uncertainty about economic conditions. In such a situation, policymakers may want to choose accessions or nudges to retention and affiliation in a way that allows the model to hit the target under as many scenarios as possible. This is an example of a class of robust decisionmaking problems, which have been extensively studied at RAND,[34] and incorporating these ideas into the model would be a useful direction for further research.

Several efforts are currently under way within the Air Force to determine the peacetime and wartime requirements for and the best mix and organization of the active and reserve components.[35] Recommendations to date have been based on manpower documents, inputs from the major commands, component end strength, force structure, basing issues, and other related considerations. None of these efforts, however, considers the effects of these decisions on the long-term health and sustainability of career fields in an integrated total force manner, as is

---

[34] See Paul K. Davis, *Analysis to Inform Defense Planning Despite Austerity*, Santa Monica, Calif.: RAND Corporation, RR-482-OSD, 2014; and Robert K. Lempert, Nidhi Kalra, Suzanne Peyraud, Zhimin Mao, Sinh Bach Tan, Dean Cira, and Alexander Lotsch, "Ensuring Robust Flood Risk Management in Ho Chi Minh City," Policy Research Working Paper No. 6465, Washington, D.C.: The World Bank, 2013.

[35] As of August 2014, AF/A1M has assembled the results of its Personnel Readiness Review, which examined the wartime requirements for the Air Force active, guard, and reserve components (excluding institutional, sustainment, and in-place requirements) to suggest those AFSCs with excess manpower that might be shifted to address shortages in other AFSCs from a total force perspective. In addition, the Total Force Continuum, under the auspices of the AF/A8, is examining component mix with respect to force structure, associations of active and reserve component units, as well as potentially shifting missions from and to the reserve components.

possible with the TFF model. We envision the TFF model as a useful tool for Air Force personnel managers for understanding historical trends and for evaluating how well future requirements can be met.

Appendix A. Transition Probabilities and Stocks and Flows

Our goal is to provide the Air Force with the ability to forecast the behavior of its military members, given plausible economic scenarios and force shaping policy alternatives. This requires understanding what we call "transition probabilities," which reflect the likely flow of service members into and out of each of the Air Force's three components.

Our approach consists of

1. Developing a simple model that relates the "state" of the total force at time t, i.e., a given fiscal year, to its state at time $(t + 1)$. Each state is a fiscal year's group of individual members in a common component who share a common entry category, a common manpower/personnel category, and a common years-of-service value.
2. Estimating the historical relationships between accessions, transition probabilities, and external economic factors, using the FY 1996-2012 time period.
3. Using these historical relationships to predict what accessions and transition probabilities will look like under different future scenarios.
4. Using the predicted accessions and probabilities, together with the model, to predict what will happen to the future total force.

To demonstrate the basic mechanics of the model, it may be instructive to examine a simple example. Imagine a total force with only two components (Active = A, and Reserve = R), only a single entry category, only a single personnel category/career field (e.g., pilots), and only a three-year career path. Pilots begin their time in the force in their first year of service, and they make choices about how to allocate their time between the two components. After three years, everyone leaves the force completely and permanently.

Let $y_{c,t}^A$ denote the number (count) of active component pilots in year-of-service cohort c at fiscal year t. Similarly, let $y_{c,t}^R$ denote the number of reserve component pilots in year-of-service cohort c at fiscal year t. We can stack these inventory counts over components and cohorts in a vector, $\boldsymbol{y}_t$, defined as follows:

$$\boldsymbol{y}_t = \begin{bmatrix} y_{1,t}^A \\ y_{2,t}^A \\ y_{3,t}^A \\ y_{1,t}^R \\ y_{2,t}^R \\ y_{3,t}^R \end{bmatrix} \qquad 1)$$

This is a (6×1) vector that stores all information about the "state" of the total force. Each element of this vector is a count of the number of pilots in each of the two components in each of the three years of service in fiscal year t.

Using some simple matrix notation, we can represent the relationship between y_t and y_{t+1} as follows:

$$y_{t+1} = M_{t+1} \, y_t + n_{t+1}.$$

Here, n_{t+1} represents a vector of gains to the total force, which come about either through accessions or rehiring of older pilots, and the matrix M_{t+1} stores all the transition probabilities. Note that the natural aging of pilots in careers, which we have limited to three years in length, sets most of the elements of this matrix to zero. In fact, there are only eight nonzero elements.

Expanding this equation in more detail, we have

$$
\begin{bmatrix}
y_{1,t+1}^A \\
y_{2,t+1}^A \\
y_{3,t+1}^A \\
y_{1,t+1}^R \\
y_{2,t+1}^R \\
y_{3,t+1}^R
\end{bmatrix}
=
\begin{bmatrix}
0 & 0 & 0 & 0 & 0 & 0 \\
M_{2,1,t+1}^A & 0 & 0 & M_{2,1,t+1}^{AR} & 0 & 0 \\
0 & M_{3,2,t+1}^A & 0 & 0 & M_{3,2,t+1}^{RA} & 0 \\
0 & 0 & 0 & 0 & 0 & 0 \\
M_{2,1,t+1}^{AR} & 0 & 0 & M_{2,1,t+1}^R & 0 & 0 \\
0 & M_{3,2,t+1}^{AR} & 0 & 0 & M_{3,2,t+1}^R & 0
\end{bmatrix}
\begin{bmatrix}
y_{1,t}^A \\
y_{2,t}^A \\
y_{3,t}^A \\
y_{1,t}^R \\
y_{2,t}^R \\
y_{3,t}^R
\end{bmatrix}
+
\begin{bmatrix}
n_{1,t+1}^A \\
n_{2,t+1}^A \\
n_{3,t+1}^A \\
n_{1,t+1}^R \\
n_{2,t+1}^R \\
n_{3,t+1}^R
\end{bmatrix}.
$$

This matrix equation represents a system of six interrelated equations. In order to better understand it, we can focus on the equation for the number of reserve component pilots with two years of service in fiscal year $t + 1$, written as $y_{2,t+1}^R$. Using simple matrix multiplication, the model gives us the following:

$$y_{2,t+1}^R = \underbrace{\left(M_{2,1,t+1}^{AR} \times y_{1,t}^A\right)}_{(A)} + \underbrace{\left(M_{2,1,t+1}^R \times y_{1,t}^R\right)}_{(B)} + \underbrace{n_{2,t+1}^R}_{(C)}.$$

This equation contains three different terms:

1. **Affiliation,** $\left(M_{2,1,t+1}^{AR} \times y_{1,t}^A\right)$: Term (A) represents the total number of pilots who decided to leave the active component after their first year and move to the reserve component. This is expressed as the total number of pilots in their first year of service, $y_{1,t}^A$, times the probability that pilots after their first year of service transition to the reserve component, $M_{2,1,t+1}^{AR}$.

2. **Aging,** $\left(M_{2,1,t+1}^R \times y_{1,t}^R\right)$: Term (B) represents the total number of pilots who spent their first year of service in the reserve component and decided to stay in the reserve

component for another year. This is expressed as the total number of reserve pilots in their first year of service, $y_{1,t}^R$, times the probability that pilots age into their next year of service in the reserve component, $M_{2,1,t+1}^R$.

3. **Total Force Prior Service Gains**, $n_{2,t+1}^R$: Term (C) represents the total number of pilots with one year of prior service who were outside the force in fiscal year t, but who are gains to the reserve component in fiscal year $t+1$. These prior service gains would be pilots who left the active component after one year of service and who transitioned back to the reserve component after a break in service.

Similar equations define the relationship between the number of pilots in each component in year of service cohorts c in fiscal year t and the number of pilots in each component in previous year of service cohorts in earlier fiscal years. While equation (2) and its matrix representation (3) may seem complex, they store all of the relationships between states together, in a compact, convenient system of linear equations that is simple to implement computationally.

Total Force Model: The Full Model

Our full total force model looks identical to the simplified example above, except that we allow for all three Air Force components (Active = A, Air Force Reserve = R, and Air National Guard = G), and we allow for 30 years of service (in addition to multiple entry categories and multiple manpower/personnel categories). After 30 years of service, we assume everyone leaves the force entirely and permanently. We can depict the model as follows:

$$\underbrace{y_{t+1}}_{(90 \times 1)} = \underbrace{M_{t+1}}_{(90 \times 90)} \underbrace{y_t}_{(90 \times 1)} + \underbrace{n_{t+1}}_{(90 \times 1)} .$$

Our previous (6×1) vectors are now (90×1), as they contain information about three components and 30 years of service, and the transition matrix, M_{t+1}, is now (90×90).

Transition Probabilities and External Factors

An innovation in our model allows accessions and transition probabilities to depend on external factors, such as the economy-wide GDP growth or the unemployment rate. We do this by specifying a simple parametric functional relationship, given by

$$M_{t+1} = M(x_{t+1}, \theta),$$

where x_{t+1} denotes a collection of time-varying measures of economic performance, and θ is a vector of parameters. We assume that this relationship, $M(x_{t+1}, \theta)$, is known.

For example, consider transitions from the Air Force Reserve. The number of possible transitions can be expressed by the set $A_R = \{R, RA, RG, RL\}$, where R denotes transitions from the Air Force Reserve to the Air Force Reserve, RA denotes transitions from the Air Force Reserve to the active component, RG denotes transitions from the Air Force Reserve to the Air

National Guard, and *RL* denotes transitions from the Air Force Reserve to losses via separation from the Air Force altogether.

To model these transitions from one cohort to the next, we assume the following:

$$M^R_{c,c-1,t+1} = \frac{\exp\{\alpha_{c,R} + x_{t+1}'\beta_c + \varepsilon_{c,R,t}\}}{1 + \sum_{s \in A_R} \exp\{\alpha_{c,s} + x_{t+1}'\beta_c + \varepsilon_{cst}\}}$$

$$M^{RA}_{c,c-1,t+1} = \frac{\exp\{\alpha_{c,RA} + x_{t+1}'\beta_c + \varepsilon_{c,RA,t}\}}{1 + \sum_{s \in A_R} \exp\{\alpha_{c,s} + x_{t+1}'\beta_c + \varepsilon_{cst}\}}$$

$$M^{RG}_{c,c-1,t+1} = \frac{\exp\{\alpha_{c,RG} + x_{t+1}'\beta_c + \varepsilon_{c,RG,t}\}}{1 + \sum_{s \in A_R} \exp\{\alpha_{c,s} + x_{t+1}'\beta_c + \varepsilon_{cst}\}}$$

$$M^{RL}_{c,c-1,t+1} = \frac{1}{1 + \sum_{s \in A_R} \exp\{\alpha_{c,s} + x_{t+1}'\beta_c + \varepsilon_{cst}\}}.$$

This functional form assumption is identical to the functional forms used in conditional logic models of discrete choice.[36] Importantly, the parameters are flexible enough to allow for significant heterogeneity. The model contains alternative specific constants, $\alpha_{c,R}$, $\alpha_{c,RA}$, and $\alpha_{c,RG}$, allowed to vary by cohort, and it contains cohort-specific slopes, β_c, which allow for heterogeneous responses of different cohorts to changes in external factors. It also allows for unobservable cohort and transition specific effects, denoted by $\varepsilon_{c,R,t}$, $\varepsilon_{c,RA,t}$, and $\varepsilon_{c,RG,t}$, to affect transition probabilities.

With the way we have specified the model's "outside option" (i.e., the probability of transitioning out of the force), we can take logs and express these transitions as linear functions of the parameters:[37]

)

) $$w^R_{c,c-1,t} \stackrel{\text{def}}{=} \ln\left(M^R_{c,c-1,t+1}\right) - \ln\left(M^{RL}_{c,c-1,t+1}\right) = \alpha_{c,R} + x_{t+1}'\beta_c + \varepsilon_{c,R,t} \cdot)$$))))

)

This means we can recover the parameters of the model from a series of linear regressions, where we regress actual transition probabilities on transition type effects and a vector of external factors, x_{t+1}.

---

[36] See, for example, Daniel McFadden, "The Measurement of Urban Travel Demand," *Journal of Public Economics*, Vol. 3, No. 4, pp. 303–328, 1974.

[37] Steven T. Berry, "Estimating Discrete-Choice Models of Product Differentiation," *RAND Journal of Economics*, Vol. 25, No. 2, 1994, pp. 242–262.

External Factor Data

To measure the strength of the economy, we used contemporaneous growth in two variables, real GDP and the federal civilian unemployment rate, as discussed in Chapter Three. These annual variables were obtained from the St. Louis Federal Reserve's FRED database. Importantly, FRED also contains projections of these annual variables out to 2015, information that was essential in our analysis.

We also used data on civilian pilot hiring from FAPA AERO. FAPA AERO is a career and financial advisory service for professional pilots and aspirants. The company specializes in providing high-quality information on labor market conditions for new and aspiring commercial pilots.

Forecasting Transition Probabilities

Let d index the initial component, and let $w_{c,t}^d$ be a vector stacking the log transformed transition probabilities from state d to all other possible states for commissioned years of service (CYOS) c at time t, as in (8):

$$w_{c,t}^d \stackrel{\text{def}}{=} \begin{bmatrix} \ln\left(M_{c,c-1,t}^{dA}\right) - \ln\left(M_{c,c-1,t}^{dL}\right) \\ \ln\left(M_{c,c-1,t}^{dR}\right) - \ln\left(M_{c,c-1,t}^{dL}\right) \\ \ln\left(M_{c,c-1,t}^{dG}\right) - \ln\left(M_{c,c-1,t}^{dL}\right) \end{bmatrix}.$$

To forecast transition probabilities under different scenarios for economic growth, we estimated regressions that take the following form:

$$w_{c,t}^d = \alpha_c^d + x_{t+1}'\beta_c + \varepsilon_{c,t},$$

where α_c^d denotes a CYOS cohort c and transition-specific fixed effect, and $\varepsilon_{c,t}$ is a mean zero error term, assumed to be uncorrelated with the regressors. We estimate this equation using a pooled panel least squares regression; separate regression models were used for each career field and commissioning source (or AFQT score group, in the case of enlisted airmen).

Table A.1 reports an example of the regression results for academy-commissioned rated pilots. Three time-varying economic covariates are included in the regression, including lagged GDP growth, unemployment rate, and total pilot hiring. All variables are interacted with an indicator for the flow direction. Because the dependent variable in this regression is a transformation of the underlying transition probabilities, the parameters of these regressions are difficult to interpret, but many of the coefficients on pilot hiring are statistically significant at the 1-percent significance level, and at least one coefficient on GDP growth is statistically significant. Importantly, the adjusted R-squared of these regressions is quite high, typically over 0.6, suggesting that the model fits the data well.

After predicting new values of $w_{c,t}^d$ under different economic scenarios and inverting the transformation, we obtain predicted transition probabilities. Figure A.1 plots the predicted and actual transition probabilities for academy pilots by commissioned years of service. There are nine graphs in this figure, and each corresponds to a separate transition (e.g. active to active, active to reserve, etc.). Actual historical transition probabilities, averaged over the 1996–2010 period, are depicted in red. Predicted transition probabilities (and 95 percent confidence bands) for FY 2015, under a low growth economic scenario, are depicted in black.

From these plots, several features are readily apparent. First, the largest transition probabilities are continuation probabilities, staying in the active, guard, or reserve. Second, the general trend of the predicted transition probabilities follows the historical averages fairly well, although confidence bands widen substantially in later years of service (when there are fewer individuals and more noise). Third, under the low economic growth scenario, we see slightly greater continuation rates than historical averages (in all three components), and lower transition probabilities of pilots from active to reserve. All features should reassure the reader in the sensibility of the model's predictions.

Table A.1: Regression Results (Example)

| | DV: w^A | DV: w^R | DV: w^G |
	(1)	(2)	(3)
gdpGrowth * (flow to A)	−0.003	0.099	0.056
	(0.048)	(0.056)*	(0.062)
gdpGrowth * (flow to R)	−0.001	−0.056	−0.095
	(0.042)	(0.058)	(0.064)
gdpGrowth * (flow to G)	−0.011	−0.022	−0.061
	(0.046)	(0.055)	(0.072)
unemp * (flow to A)	−0.029	−0.072	−0.004
	(0.111)	(0.114)	(0.109)
unemp * (flow to R)	−0.088	−0.138	0.014
	(0.080)	(0.102)	(0.117)
unemp * (flow to G)	−0.084	−0.106	−0.011
	(0.081)	(0.102)	(0.103)
pilotHiring * (flow to A)	0.000	0.000	0.000
	(0.000)**	(0.000)***	(0.000)***
pilotHiring * (flow to R)	0.000	0.000	0.000
	(0.000)**	(0.000)*	(0.000)
pilotHiring * (flow to G)	0.000	0.000	0.000
	(0.000)***	(0.000)*	(0.000)
Adjusted R-Squared	0.817	0.728	0.676
N	1260	1260	1260
F-Stat	281.49	397.85	199.76
CYOS x Dest. Compo FE	Yes	Yes	Yes

Note: This table reports regression results from estimates of equation (9) for Rated Pilots from the Air Force Academy. Each column uses a different dependent variable. All regressions include CYOS by destination component fixed effects. Robust standard errors are reported in parentheses. */**/*** denotes significant at the 90/95/99-percent significance level.

Figure A.1: Predicted and Actual Transition Probabilities (Example)

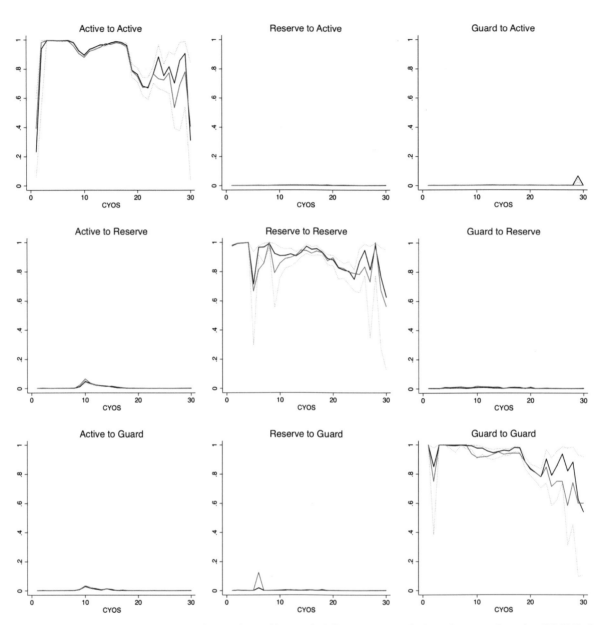

NOTE: This figure depicts predicted and actual transition probabilities, by commissioned years of service (CYOS), for rated pilots from the academy commissioning source. Actual transition probabilities, averaged over the 1996–2010 period, are depicted in red. Predicted transition probabilities under a low-growth economic scenario (and confidence bands) for FY 2015 are depicted in black.

Appendix B. Mathematical Formulation of the Optimization Approach

This appendix outlines the mathematical formulation of the new optimization approach described in Chapter Four. The basic model is a linear program, in which a hierarchical series of objective functions is considered using sequential optimizations. Specifically, the model contains a series of three consecutive objectives where each objective is used as a constraint in subsequent objectives.

- The first objective constrains the aggregate number of individuals that must be present in each component over time based on the initial inventory.

- The second objective attempts to achieve the required experience distribution in each component, while maintaining the first objective's aggregate number of individuals in each component.

- The third objective attempts to minimize the number of changes to baseline transition and accession rates required to adhere to the results of the first two objectives.

Given that we currently have minimal insight into the cost structure associated with making changes to the baseline transitions and accessions, these objective functions were structured in an effort to minimize the number of personnel transitions. Calculating the associated costs will require more research and analyses.

First Objective Function

The goal of the first objective is to minimize the deviations between the total individuals in each component, regardless of experience requirements, and the total requirement for individuals in each component over all the future years under consideration. The result of the objective is used to constrain the aggregate number of individuals that must be present in each component over time for future objectives. As an input to the model, this objective operates on an initial inventory, baseline accessions of individuals into each *state*, and baseline transition probabilities from each *state* to other *states*.[38] The model is then allowed to make adjustments from these baseline accession and transition rates in accordance with the first objective; however, these deviations are constrained in a number of important ways:

- The model can only adjust accessions to a value between zero accessions and a maximum number of accessions that is an input into the model. As a simplifying assumption, the model bounds only the number of experienced individuals (four years of service or more)

---

[38] As a reminder, each state is a fiscal year's group of individuals in a common component who share a common entry category, a common manpower/personnel category, and a common year-of-service value.

and inexperienced individuals (three years of service or less) that can be assessed into each component during each year, as opposed to placing a bound on each potential *state*.

- The model can alter only the *state* each individual transitions into, not the number of individuals that transition. For example, assume 25 individuals are transitioning from the active component with ten years of service to the reserve component with 11 years of service. The model could adjust this to be 15 individuals transitioning into the reserve component with 11 years of service and 10 individuals transitioning into the active component with 11 years of service.

- As an input to the model, an upper bound is placed on the total number of experienced individuals who can transition from one component to a different component each year.

- Finally, the model does not allow deviations to the baseline transitions into states associated with inexperienced individuals.

Given these constraints, the model attempts to move the initial inventory to match the aggregate requirement for each component over all future years. It is important to point out that this may yield large deviations in single years, because the goal is to minimize the discrepancy over time, rather than the deviations for a single year. The result of this optimization step is a bound on the total number of individuals that should be present in each component for each future year. This bound is then used as an input for the second optimization step.

Second Objective Function

Having brought each component's total inventory as close as possible to its aggregate requirement over all the future years under consideration, the second objective accounts for differences in experience, as measured by years of service, attempting to achieve the required experience distribution of individuals in each component. The second objective, then, solves a new optimization problem that minimizes the sum of the absolute deviations between the number of individuals in each *state* and the requirement for individuals in that *state,* over all years. The main constraint to the second objective is that the only feasible solutions are those that maintain the aggregate number of individuals in each component within the bounds determined by the first objective. The results of this optimization, combined with the results of the first optimization, are then used as bounds in the third and final optimization step.

Third Objective Function

As mentioned previously, we currently have little insight into the cost structure associated with making changes to the baseline accession and transition rates. While, for example, there is likely some cost to the Air Force to induce a higher-than-expected number of individuals at a given year of service to remain in the force (or to depart the force), we know neither the magnitude of the costs nor the nature of the relationship between such actions and costs, such as whether there are both fixed and variable components to these costs. Subsequent analysis could

60

identify such cost structures, at which point it will be relatively straightforward to incorporate into this modeling framework a new objective function that minimizes the cost of attaining the required inventory.

Because we did not have such cost information, the third objective instead utilizes an approach that minimizes the sum of the absolute deviations from the baseline estimates for accessions and transitions. As was done for the first objective, the model similarly needs to bound the second objective function's value, to maintain the count of individuals in each state within some defined error bound. The result of the third objective is the minimized sum of the absolute deviations from the baseline estimates for accessions and transitions over all years, while maintaining the results from the first two objectives within a given error bound.

First Objective Details

To set up the first objective, we begin by defining a set I consisting of all possible states. For example, one possible state is rated officers commissioned by ROTC and currently in the active component with five years of service. For convenience of notation, we add an element i_0 to set I to denote individuals who depart the total force. We also define a set T consisting of all years under consideration in the analysis.

We track the inventory of individuals in each state through the following elements:

Input parameter L_i In tt e Inltlal Inventory of Indlvldualn In ntate III II

I eplnlon varlable $L_{t,i}$ In tt e Inventory of Indlvldualn In ntate L at tt e end of flnpal year L I t ere $Y_{t,i} \geq 0$ I

The model represents the accession and transition decisions as potential deviations from the baseline estimates of accessions and transitions developed utilizing the econometric analysis presented in Chapter Four. The model defines the following elements:

Input parameter $M_{t,i,\hat{i}}$ In tt e banellne trannltlon probablllty from ntate L to ntate $\hat{i}$ durln flnpal year II

Input parameter $N_{t,i}$ In tt e banellne appennlonn of Indlvldualn Into ntate L durln flnpal year II

I eplnlon varlable $X_{t,i,\hat{i}}$ In tt e devlatlon from tt e banellne trannltlon of Indlvldualn from ntate L to ntate $\hat{i}$ durln flnpal year II

I eplnlon varlable $Z_{t,i}$ In tt e devlatlon from tt e banellne appennlonn $N_{t,i}$ of Indlvldualn Into ntate II durln flnpal year II

The model tracks the impact of accession and transition decisions on the inventory through a set of flow balancing constraints:

$$Y_{"t=1",i} = \sum_{\hat{\imath}} \left(M_{"t=1",\hat{\imath},i} * L_{\hat{\imath}} + X_{"t=1",\hat{\imath},i} \right) + N_{"t=1",i} + Z_{"t=1",i} \qquad \forall i \geq 1 \tag{1}$$

$$Y_{t,i} = \sum_{\hat{\imath}} \left(M_{t,\hat{\imath},i} * Y_{t-1,\hat{\imath}} + X_{t,\hat{\imath},i} \right) + N_{t,i} + Z_{t,i} \qquad \forall i \geq 1; t \geq 2 . \tag{2}$$

Using the initial inventory that was input into the model, Constraint (1) calculates the new inventory for each state after the first year. Specifically, the constraint calculates the adjusted baseline transitions from the initial inventory states and the adjusted baseline accessions into each state for the first year. Constraint (2) calculates the same for each year after the first using the inventory result computed for the prior year by the model (e.g., year 2 is computed using the new inventory determined for year 1).

Additional constraints are necessary to ensure that the positive-value $X_{t,i,\hat{\imath}}$ deviations do not exceed the total number of individuals who could potentially be redirected out of state i during year t:

$$X_{"t=1",i,\hat{\imath}} \leq \left(1 - M_{"t=1",i,\hat{\imath}} \right) * L_i \qquad \forall i \geq 1 \tag{3}$$

$$X_{t,i,\hat{\imath}} \leq \left(1 - M_{t,i,\hat{\imath}} \right) * Y_{t-1,i} \qquad \forall i \geq 1; t \geq 2 \tag{4}$$

Coupled with Constraint (1), Constraint (3) ensures that the baseline transition of individuals, along with the deviation to those transitions, cannot exceed the total initial inventory available. Constraint (4) ensures the same for each consecutive year based on the new inventory calculated for the previous year.

Other constraints are necessary to ensure that the negative-value $X_{t,i,\hat{\imath}}$ deviations do not exceed the total number of individuals who could potentially be redirected out of state i during year t:

$$X_{"t=1",i,\hat{\imath}} \geq -M_{"t=1",i,\hat{\imath}} * L_i \qquad \forall i \geq 1 \tag{5}$$

$$X_{t,i,\hat{\imath}} \geq -M_{t,i,\hat{\imath}} * Y_{t-1,i} \qquad \forall i \geq 1; t \geq 2 . \tag{6}$$

A negative deviation (i.e., $X_{t,i,\hat{\imath}} < 0$) to the transition of individuals prevents individuals from leaving state i and transitioning into state $\hat{\imath}$. Constraints (5) and (6) are necessary to ensure that the negative deviation to the transition of individuals out of state i and into state $\hat{\imath}$ cannot actually introduce new individuals into the original state i. Specifically, this deviation can only reduce the number of individuals transitioning out of state i to 0 and not less than 0, which would imply the transition of individuals into state i.

Another constraint is necessary to ensure that the sum total of $X_{t,i,\hat{\imath}}$ deviations out of state i during year t equals zero:

$$\sum_{\hat{i}} X_{t,i,\hat{i}} = 0 \quad \forall t; \ i \geq 1.\tag{7}$$

Decision variable $X_{t,i,\hat{i}}$ is allowed to change only the states $\hat{i}$ individuals are transitioning to from i, not the total number of individuals transitioning out of state i. Constraint (7) ensures that the total number of individuals transitioning from state i remains constant (i.e., if more individuals are transitioning into state i' from i, then an equivalent amount of individuals must be transitioning into at least one or more other states i'' from i).

Another constraint is necessary to ensure that the negative-value $Z_{t,i}$ deviations to accessions do not exceed the total number of accessions into state i during year t that could potentially be eliminated:

$$Z_{t,i} \geq -N_{t,i} \quad \forall t; \ i \geq 1.\tag{8}$$

The model imposes upper bounds on accessions and further constrains transitions, as follows. First, the model defines the following subsets of I, the set of all possible states:

subset I_c, denoting those elements of set I that correspond to component c

subset I_{exp}, denoting those elements of set I that correspond to experienced individuals

subset I_{inexp}, denoting those elements of set I that correspond to inexperienced individuals.

Now the model defines

input parameter $P_{c,t}$ as an upper bound on the total number of experienced individuals who can be accessed into component c from outside the total force during fiscal year t

input parameter $Q_{c,t}$ as an upper bound on the total number of inexperienced individuals who can be accessed into component c from outside the total force during fiscal year t

input parameter $S_{c,\hat{c},t}$ as an upper bound on the total number of experienced individuals who can transition from component c to component $\hat{c} \neq c$ during fiscal year t.

The following constraints enforce these upper bounds:

$$\sum_{i \in I_c \cap I_{exp}} \left(Z_{t,i} + N_{t,i} \right) \leq P_{c,t} \quad \forall c, t\tag{9}$$

$$\sum_{i \in I_c \cap I_{inexp}} \left(Z_{t,i} + N_{t,i} \right) \leq Q_{c,t} \quad \forall c, t\tag{10}$$

$$\sum_{\substack{i \in I_c \\ \hat{i} \in I_{\hat{c}} \cap I_{exp}; \hat{c} \neq c}} \left(X_{"t=1",i,\hat{i}} + M_{"t=1",i,\hat{i}} * L_i \right) \leq S_{c,\hat{c},"t=1"} \quad \forall c, \hat{c} \neq c \tag{11}$$

$$\sum_{\substack{i \in I_c \\ \hat{i} \in I_{\hat{c}} \cap I_{exp}; \hat{c} \neq c}} \left(X_{t,i,\hat{i}} + M_{t,i,\hat{i}} * Y_{t-1,i} \right) \leq S_{c,\hat{c},t} \quad \forall c, \hat{c} \neq c; t \geq 2 \,. \tag{12}$$

The model does not allow deviations to the baseline transitions into states associated with inexperienced individuals:

$$X_{t,i,\hat{i}} = 0 \quad \forall t, i, \hat{i} \in I_{inexp} \,. \tag{13}$$

Note that the model also allows transitions to occur only from a state associated with years of service = a to either a state with years of service = $a + 1$, or the "sink" state i_0 for individuals who depart the total force. To ensure this, the model sets $M_{t,i,\hat{i}} = 0$ for all cases where the years of service for $\hat{i} \neq$ years of service for $i + 1$.

The model now compares inventories to requirements using the following elements:

Input parameter $R_{t,i}$ is the requirement for individuals in state i at the end of fiscal year t.

Decision variable $V_{t,c}$ is the deviation from the aggregate requirement for component c at the end of fiscal year t.

The aggregate requirement constraint to determine $V_{t,c}$ for each component c in each fiscal year t can be written as

$$\sum_{i \in I_c} (Y_{t,i} - R_{t,i}) = V_{t,c} \quad \forall t, c \,. \tag{14}$$

In order to allow the model to track the absolute value of the deviations from the aggregate requirements, we employ a standard modeling technique, introducing new variables:

Decision variable $Vplus_{t,c}$ is the absolute value of $V_{t,c}$ if $V_{t,c}$ is positive; $Vplus_{t,c} \geq 0$.

Decision variable $Vminus_{t,c}$ is the absolute value of $V_{t,c}$ if $V_{t,c}$ is negative; $Vminus_{t,c} \geq 0$.

The model then needs constraint:

$$V_{t,c} = Vplus_{t,c} - Vminus_{t,c} \quad \forall t, c \,. \tag{15}$$

The first objective is to get the total number of individuals in each component as close as possible to that component's aggregate requirement, making no distinction between individuals with different year-of-service values within any component. To do this, we solve an optimization problem that minimizes obj_1, the sum of $|V_{t,c}|$ over all components c and all fiscal years t (define α as the optimal value for this objective function):

$$obj_1 = \sum_{t,c} \left(Vplus_{t,c} + Vminus_{t,c} \right). \tag{16}$$

Second Objective Details

Having brought each component's total inventory as close as possible to its aggregate requirement, the second objective accounts for year-of-service differences, attempting to produce the required years-of-service distribution of individuals in each component, while maintaining the previous optimization's aggregate number of individuals in each component. To set up the second objective, the model defines the following additional variables:

Decision variable $W_{t,i}$ is the deviation from requirement $R_{t,i}$.

Decision variable $Wplus_{t,i}$ is the absolute value of $W_{t,i}$ if $W_{t,i}$ is positive; $Wplus_{t,i} \geq 0$.

Decision variable $Wminus_{t,i}$ is the absolute value of $W_{t,i}$ if $W_{t,i}$ is negative; $Wminus_{t,i} \geq 0$.

The model then needs constraints:

$$Y_{t,i} - R_{t,i} = W_{t,i} \qquad \forall t, i \geq 1 \tag{17}$$
$$W_{t,i} = Wplus_{t,i} - Wminus_{t,i} \qquad \forall t, i \geq 1. \tag{18}$$

The model also needs to bound the first objective function's value, to maintain the aggregate count of individuals in each component, which we allowed within a range of $\pm 0.5\%$):

$$obj_1 \geq 0.995 * \alpha \tag{19}$$
$$obj_1 \leq 1.005 * \alpha. \tag{20}$$

Constraints (19) and (20) ensure that the only feasible solutions to the second objective are solutions that maintain the total number of individuals in each component within some defined error bound of the total number of individuals in each component determined by the first objective. The second objective, then, solves a new optimization problem that minimizes obj_2, the sum of the absolute deviations between the number of individuals in each state i and the requirement for individuals in state i, over all fiscal years t (define β as the optimal value for this objective function):

65

$$obj_2 = \sum_{t,i} \left(Wplus_{t,i} + Wminus_{t,i} \right).$$ (21)

Third Objective Details

The third objective utilizes an approach that minimizes the sum of the absolute deviations from the baseline estimates for accessions and transitions as a proxy for cost. To do this, the model defines a new set of variables to accommodate absolute values:

Decision variable $Xplus_{t,i,\hat{\imath}}$ is the absolute value of $X_{t,i,\hat{\imath}}$ if $X_{t,i,\hat{\imath}}$ is positive;
$Xplus_{t,i,\hat{\imath}} \geq 0$.

Decision variable $Xminus_{t,i,\hat{\imath}}$ is the absolute value of $X_{t,i,\hat{\imath}}$ if $X_{t,i,\hat{\imath}}$ is negative;
$Xminus_{t,i,\hat{\imath}} \geq 0$.

Decision variable $Zplus_{t,i}$ is the absolute value of $Z_{t,i}$ if $Z_{t,i}$ is positive;
$Zplus_{t,i} \geq 0$.

Decision variable $Zminus_{t,i}$ is the absolute value of $Z_{t,i}$ if $Z_{t,i}$ is negative;
$Zminus_{t,i} \geq 0$.

The model then needs constraints:

$$X_{t,i,\hat{\imath}} = Xplus_{t,i,\hat{\imath}} - Xminus_{t,i,\hat{\imath}} \qquad \forall t, i, \hat{\imath}$$ (22)

$$Z_{t,i} = Zplus_{t,i} - Zminus_{t,i} \qquad \forall t, i .$$ (23)

As was done for the first objective, the model similarly needs to bound the second objective function's value, to maintain the count of individuals in each state, which we allowed within a range of $\pm 0.5\%$):

$$obj_2 \geq 0.995 * \beta$$ (24)

$$obj_2 \leq 1.005 * \beta .$$ (25)

The third objective solves a new optimization problem that minimizes obj_3, the sum of the absolute deviations from the baseline estimates for accessions and transitions, over all fiscal years t, while maintaining the results from the first two objectives within a given error bound:

$$obj_3 = \sum_{t,i,\hat{\imath}} \left(Xplus_{t,i,\hat{\imath}} + Xminus_{t,i,\hat{\imath}} \right) + \sum_{t,i} \left(Zplus_{t,i} + Zminus_{t,i} \right).$$ (26)

References

Air Force Audit Agency, *Air Force Personnel Reductions—Audit Report F2008-0004-FD4000*, May 12, 2008.

AFI 36-2604, *Service Dates and Dates of Rank*, October 5, 2012.

Asch, Beth J., James R. Hosek, Michael G. Mattock, and Christina Panis, *Assessing Compensation Reform: Research in Support of the 10th Quadrennial Review of Military Compensation*, Santa Monica, Calif.: RAND Corporation, MG-764-OSD, 2008. As of October 1, 2015:
http://www.rand.org/pubs/monographs/MG764.html

Asch, Beth J., Michael G. Mattock, and James R. Hosek, *A New Tool for Assessing Workforce Management Policies Over Time: Extending the Dynamic Retention Model*, Santa Monica, Calif.: RAND Corporation, RR-113-OSD, 2013. As of October 1, 2015:
http://www.rand.org/pubs/research_reports/RR113.html

Berry, Steven T., "Estimating Discrete-Choice Models of Product Differentiation," *RAND Journal of Economics,* Vol. 25, No. 2, 1994, pp. 242–262.

Binkin, Martin, *Who Will Fight the Next War? The Changing Face of the American Military*, Washington, D.C.: The Brookings Institution, 1993.

Davis, Paul K., *Analysis to Inform Defense Planning Despite Austerity*, Santa Monica, Calif.: RAND Corporation, RR-482-OSD, 2014. As of October 1, 2015:
http://www.rand.org/pubs/research_reports/RR482.html

Department of Defense, DoD 7000.14-R, *Financial Management Regulation*, Volume 7A, April 2013.

James, Deborah Lee, and Mark A. Welsh III, Written Statement to the Senate, Committee on Armed Services, *The National Commission on the Structure of the Air Force*, April 29, 2014. As of October 1, 2015:
http://www.armed-services.senate.gov/imo/media/doc/14-42%20-%204-29-14.pdf

Lempert, Robert, Nidhi Kalra, Suzanne Peyraud, Zhimin Mao, Sinh Bach Tan, Dean Cira, and Alexander Lotsch, "Ensuring Robust Flood Risk Management in Ho Chi Minh City," Policy Research Working Paper No. 6465, Washington, D.C.: The World Bank, 2013. As of October 1, 2015:
http://www.rand.org/pubs/external_publications/EP50282.html

McFadden, Daniel, "The Measurement of Urban Travel Demand," *Journal of Public Economics*, Vol. 3, No. 4, pp. 303–328, 1974.

National Commission on the Structure of the Air Force, *Report to the President and Congress of the United States*, Arlington, Va., January 30, 2014. As of October 1, 2015: http://afcommission.whs.mil/public/docs/NCSAF%20WEB220.pdf

Robbert, Albert A., Lisa M. Harrington, Tara L. Terry, and Hugh G. Massey, *Air Force Manpower Requirements and Component Mix: A Focus On Agile Combat Support*, Santa Monica, Calif.: RAND Corporation, RR-617-AF, 2014. As of October 1, 2015: http://www.rand.org/pubs/research_reports/RR617.html

Rostker, Bernard D., Charles Robert Roll, Jr., Marney Peet, Marygail K. Brauner, Harry J. Thie, Roger Allen Brown, Glenn A. Gotz, Steve Drezner, Bruce W. Don, Ken Watman, Michael G. Shanley ,Fred L. Frostic, Colin O. Halvorson, Norman T. O'Meara, Jeanne M. Jarvaise, Robert Howe, David A. Shlapak, William Schwabe, Adele R. Palmer, James H. Bigelow, Joseph G. Bolten, Deena Dizengoff, Jennifer H. Kawata, Hugh G. Massey, Robert Petruschell, S. Craig Moore, Thomas F. Lippiatt, Ronald E. Sortor, J. Michael Polich, David W. Grissmer, Sheila Nataraj Kirby, and Richard Buddin, *Assessing the Structure and Mix of Future Active and Reserve Forces*, Santa Monica, Calif.: RAND Corporation, MR-140-1-OSD, 1992. As of October 1, 2015: http://www.rand.org/pubs/monograph_reports/MR140-1.html

Abbreviations

AFPC	Air Force Personnel Center
AFQT	Armed Forces Qualifying Test
AFSC	Air Force Specialty Code
ABM	air battle manager
CAFSC	control AFSC
CCR	cumulative continuation rate
CEA	career enlisted aviator
CEM	Chief Enlisted Manager
CSO	combat systems officer
DIEUS	Date Initial Entry Uniformed Service
DoD	Department of Defense
EAD	entry on active duty
FRED	Federal Reserve Economic Data
GDP	gross domestic product
MPES	[Air Force] Manpower Programming and Execution System
NPS	nonprior service
PAA	Primary Aircraft Authorization
PAF	Project AIR FORCE
PBD	Program Budget Directive
PP	permanent party
RDTM	rated distribution and training management
ROTC	Reserve Officer Training Corps
RPA	remotely piloted aircraft
SAF/MR	Assistant Secretary of the Air Force for Manpower and Reserve Affairs
SAS	Statistical Analysis Software

STP	students, transients, and personnel holdees
TFBL	Total Force Blue Line
TFCSD	Total Federal Commissioned Service Date
TFF	Total Force Flow Model
UMD	unit manpower documents